CURRENT

SHOCKED

David Casarett, MD, MA, is a physician, researcher, and tenured professor at the University of Pennsylvania Perelman School of Medicine. His studies have included more than ten thousand patients and have resulted in more than one hundred articles and book chapters, published in leading medical journals such as the *Journal of the American Medical Association* and the *New England Journal of Medicine*. His many awards include the prestigious U.S. Presidential Early Career Award for Scientists and Engineers. He lives in Philadelphia.

It is the heart that kills us in the end
Just one more old broken bone that cannot mend

—Emmylou Harris, "The Pearl"

SHOCKED

A Doctor Investigates the Blurred Lines Between Life and Death

DAVID CASARETT, MD

Current

CURRENT
An imprint of Penguin Random House LLC
375 Hudson Street
New York, New York 10014
penguin.com

First published in the United States of America by Current 2014
This paperback edition published 2015

THE LIBRARY OF CONGRESS HAS CATALOGED THE HARDCOVER EDITION AS FOLLOWS:

Casarett, David J., author.
Shocked : adventures in bringing back the recently dead / David Casarett.
p. cm.
Includes bibliographical references and index.
ISBN 978-1-59184-671-0 (hc.)
ISBN 978-1-61723-022-6 (pbk.)
I. Title.
[DNLM: 1. Resuscitation—Popular Works. WA 292]
RC86.7
616.02'5—dc23
2014004313

Printed in the United States of America
10 9 8 7 6 5 4 3 2 1

Set in Adobe Caslon Pro
Designed by Spring Hoteling

CONTENTS

1

The Big Mac Rule of Resuscitation and the Search for the Limits of Life

When I was a kid, long before I contemplated going to medical school, the television in our living room was the sole source of all of my medical knowledge. Before I ever dissected a cadaver or listened to a heart, shows like *M*A*S*H*; *St. Elsewhere*; *Doogie Howser, MD*; *Chicago Hope*; and *ER* taught me how to be a doctor. Specifically, they taught me that doctors are firm, decisive, quick-thinking, and almost always successful.

Television also taught me how to bring someone back to life. Fortunately, that was a simple lesson for an eight-year-old. The television version of resuscitation followed a script that was mercifully predictable, and that predictability was helpfully marked by several reliable guideposts along the way.

First, someone's heart would stop. That cessation of a heartbeat was usually heralded by unmistakable signs, including but not limited to gasping, choking, eye rolling, and chest clutching.

Next, and typically without any discernible delay whatsoever, everyone within hailing distance would descend on the newly dead character. One of these self-appointed rescuers would then place two hands on the character's chest and bounce up and down heroically. It was also at about this point that another rescuer—usually a tall, handsome doctor—performed a strange sort of kissy procedure with his mouth, guaranteed to provoke slack-jawed fascination in a boy not yet in middle school, especially if the victim was a woman. Finally, if the episode were really top-notch, someone would produce a pair of paddles, apply them to the victim's chest, and yell, "Clear!" (At some point, I developed the unshakable conviction that this shouted incantation had some ill-defined yet essential electrical effect on the victim's heart. I have a hazy recollection of standing over my freshly late hamster one sad morning and yelling, "Clear!" repeatedly in hopes of encouraging little Frankie to rejoin the living. Alas, Frankie was unfamiliar with the rules of televised resuscitations, and he remained persistently and unambiguously deceased.)

Then there would be a strategic yet wholly incongruous commercial break, after which we'd be back in the thick of things. On cue, the victim would tire of being kissed by a tall, handsome doctor and would wake up. Or, occasionally—and just for variety's sake—the handsome doctor would tire of kissing a person who was becoming increasingly dead. Then he would stand up, say something solemn, and stride off purposefully toward the next crisis.

It was thanks to these scenes that I developed a deep and lasting impression of how resuscitation works when people try to die. For instance, I came to believe that resuscitation *works*. Maybe not always, but almost always. It seemed as though even if you were dead, as long as there was a good-looking doctor nearby, you wouldn't be dead for long.

I also became convinced that if resuscitation is going to work, it's going to work very, very fast. A perceptive watcher of these shows would conclude that the fate of a newly dead person is determined in the span of time that it takes to learn about the merits of cookies made by Keebler Elves or a sing-along of the McDonald's Big Mac jingle. Let's call this the Big Mac rule of resuscitation. By then, your victim is probably wide-awake and hugging the rescuers. If she isn't, then you might as well switch channels.

So I persisted in my fantasies about resuscitation for quite some time.

But then a girl named Michelle died.

THE MIRACLE GIRL

Years later, I was driving home from college when I stopped at a rest area on the Pennsylvania Turnpike for a bite to eat. As I took a seat I found a weathered copy of a local paper stuck to the table by layers of French fry grease and ice cream scum. Amid reports of troubles in city government and announcements of tax hikes and education cuts, one story caught my attention.

A two-and-a-half-year-old girl named Michelle Funk had fallen into a stream and drowned a week or so earlier. By the time rescuers pulled her out—more than an hour after the accident—she was dead. Not just dead by the Big Mac rule, but really dead. Really, truly dead.

Then the article went on to report that she was alive. In critical condition, but alive.

Huh?

I reread that passage a couple of times. Michelle Funk had really died. And then she really was alive. So much so that she was on her way home.

The full story, which I read about much later in a medical journal, was even more impressive. On June 10, 1986, in Salt Lake City, Michelle was playing with her brother near a creek still swollen with snowmelt when she slipped and fell in. Her brother couldn't help her, so he ran to get their mother, who searched frantically for several minutes before calling 911.

Paramedics arrived quickly, but by the time they were able find her and bring her to the creek bank, sixty or so minutes had passed. For more than an hour Michelle hadn't been breathing. And for probably almost as long, her heart hadn't been beating.

As she lay on the bank of the creek, the paramedics could see that Michelle was cold and lifeless. The journal article describing her case put her condition in stark medical terms: "The child was cyanotic, apneic, and flaccid," it reports, "with fixed and dilated pupils and no palpable pulse."

In English, that means that she was dead. She wasn't breathing (that is, she was *apneic*). She also had the dusky-blue (*cyanotic*) color of someone whose body has been starved of oxygen. And the fact that her pupils were fixed and dilated meant that her brain—specifically her brain stem, one of the areas that controls the eyes' response to light—had shut down.

Put yourself in the position of those paramedics, standing next to Michelle on the bank of the creek. You have to decide whether to try to resuscitate her. What would you do?

Would you walk away? You might.

But you might also think: She's very young. And isn't even a tiny chance of success worth it for a two-year-old?

At the time, no one had ever survived after more than an hour underwater. In fact, the conventional wisdom was that drowning victims like Michelle had the best chance of survival if they were revived within fifteen minutes. That chance dropped precipitously after a victim had been underwater more than twenty minutes. And Michelle had been in the water for an hour.

Now would you walk away?

Logically, you probably should. And most paramedics probably would have. There was no reason to believe that they could do any good. There was no point in trying.

But those paramedics did try to revive Michelle Funk. For some reason—intuition, instinct, or just blind hope—they thought that they might be able to bring her back.

So they transported Michelle to an emergency room, where she was met by a team that began to try to get her heart beating. They tried all of the tricks they could think of, without success. One hour turned into two. Then two hours turned into three.

Then Michelle took a small, almost undetectable breath. A moment later, her heart began to show evidence of activity. This was only a confused flutter, the medical term being *fibrillation*—little more than background noise. But soon it became more organized, breaking spontaneously into a normal rhythm. Then she was alive. After three hours of being dead, Michelle Funk was alive.

Later, when Michelle's survival was described in the prestigious *Journal of the American Medical Association*, an editorial accompanying that article called Michelle's survival "miraculous."

Try searching the pages of a mainstream medical journal for that word, and you'll find that it's rare indeed.

Michelle's story left a deep impression on me, and I decided right then, sitting in that rest stop on the Pennsylvania Turnpike, that I wanted to be an emergency room physician. I wanted to bring people like Michelle back to life. And I wanted to see just how far I could push science. If three hours was possible, what about twelve? Or a day? Or a month?

For the rest of my six-hour drive home, I kept thinking about Michelle. One day, I thought, I'd like to find out what happened to her.

Twenty years later, I did, and I found out enough to draw my own conclusions about whether cutting-edge resuscitation for someone like Michelle Funk is a good thing.

A LESSON IN BASS FISHING

Michelle's story seemed clear-cut and almost cinematic in its simplicity when I first encountered it. She was a little girl who was dead, and then she was alive. That really was worth celebrating, and it struck me as fair for that journal article to have called her survival "miraculous." But a few years later, in medical school, I met a patient who taught me that these sorts of stories are rare.

Joe was one of the first patients I ever took care of as a medical student. He was a gregarious guy. A retired steelworker, Joe loved bass fishing more than anything else in the world. He'd happily display a Polaroid gallery of fish he'd caught to anyone who evinced even a modest interest.

Joe was admitted to the hospital with an almost complete blockage of his left main coronary artery, the principal supply of blood to most of the heart. That blockage had been discovered during a cardiac catheterization that morning. When I met him, Joe was waiting for surgery that would take place the same afternoon. His blockage was severe, and we knew it would only take a little exertion on his part for his heart to shut down entirely.

Joe and I were sitting in his hospital room, and he'd just told me that the secret to bass fishing was quiet. You had to sneak up on them, he said. Especially the big ones. They couldn't see worth a damn but their hearing, he claimed, was acute.

Joe was propped up in bed and I was perched on the arm of a chair nearby. I was only half-listening. Instead, most of my attention was focused on the chart on my lap, open to the page that described the results of Joe's cardiac catheterization.

Then I noticed that Joe wasn't talking anymore. I thought that perhaps he was illustrating the principle of quiet. He was showing me how a successful fisherman sneaks up on an unsuspecting bass. I smiled and I looked up.

As I did, I saw Joe staring at me with the vacant expression of alarm that you might wear if the person in front of you has a spider on his forehead. Then Joe's eyes rolled back in his head, and he took a couple of gasping breaths. Then he stopped moving.

My medical training kicked in and I checked for a pulse. Then I ran to the hall and asked a nurse to call a code—medical-speak for gathering doctors and nurses from all over the hospital to try to resuscitate a patient who has had a cardiac arrest. In the meantime, another nurse and I started CPR.

A minute went by. Then two. Those two minutes were about the longest two minutes I've ever experienced.

Finally, the resuscitation team arrived, and I was grateful to step back into a supporting role. I told the resident what I knew about Joe, and his left main blockage. Then the code team ran through the drill of defibrillation and injections of medications.

After about fifteen minutes, they got a heart rhythm back. By that time the team had inserted an endotracheal tube into Joe's lungs and was breathing for him. He had a heart rate and blood pressure, which was good.

However, an EKG showed that he'd had a massive heart attack, which was very bad. The resident had Joe's cardiothoracic surgeon on the phone, and he decided that Joe needed to go to the operating room for surgery right then. So off they went, leaving behind an empty bed and a pile of well-thumbed fishing magazines.

Joe made it through surgery, technically, but he never woke up. His heart was too badly damaged to pump blood effectively, and the rest of him had "taken a beating" too, as Joe would have said. Both his liver and his kidneys had failed, for instance. And his brain had probably been severely damaged by being without oxygen for those fifteen long minutes while we were performing CPR.

I went to see him every day in the ICU. I looked at his chart. I checked his lab tests. And I looked at Joe.

Every day I saw his family—wife, three grown children, half a dozen grandchildren—filing in and out of his room, having tearful conversations with the surgeon. They'd be hopeful one day as they got some good news, and then sad the next. That went on for eighteen long days.

And every one of those eighteen days I wondered whether I should have been so quick to run out into the hall to call the code and begin

CPR right away. I did everything by the book, like a good Boy Scout. But by my logic, thanks to me, Joe and his family were stuck in a medical limbo that they couldn't escape.

On the nineteenth day, Joe's bed was occupied by a young woman who'd barely survived a motorcycle accident. I learned that the previous evening the surgeons had said that there was nothing else they could do for Joe. His family was sad, but—I imagine—a little relieved, too. They agreed to take him off the ventilator. He wasn't able to breathe on his own, and so he died quickly.

THE TITHONUS PROBLEM

It was as a result of my experience with Joe that I've strayed a little in my medical career. Well, more than a little. I started medical school wanting to work in an emergency room, but now I'm a hospice doctor. It's a little like trying out for the Philadelphia Eagles as a quarterback and ending up as the coach. Not better or worse, but very, very different.

At least once a week I see a patient who makes me think about either Michelle or Joe, or both. Some days, I see miracles like Michelle. These are people who have benefited from advances in the science of resuscitation and who have gained years of good life. And each time I meet one of these patients, I wonder what's possible. I think about the science that will be available to all of us in a year, or in a decade. And I wonder whether Michelle's miraculous survival is the tip of the iceberg in terms of preserving human life.

But other days, I meet a patient like Joe. These are people who have been kept alive by the same advances, often for weeks or months in an ICU. When I meet those patients, I have to ask whether we should be trying so hard to push the limits of what's possible. Maybe we should heed the Big Mac rule? Maybe if someone can't be revived quickly and easily, we should leave well enough alone?

Most of all, I wonder how technology is going to change the way that we die. Because if there's one thing I've learned as a hospice doctor, it's that we're all going to die. Sure, the science of resuscitation can delay death for minutes or hours or—for someone like Michelle—decades. But it's still going to happen to all of us.

Although the science of resurrection can rescue people like Michelle, those rescues come at a cost. What kinds of costs? Well, since Michelle's story achieved almost mythical status in popular culture, it makes sense to look for the answer to that question in mythology.

Tithonus is described in Greek mythology as a mortal, one of the lovers of Eos, goddess of the dawn. That, you'd think, would be enough to make any guy happy. Who could ask for more?

Well, he did ask for more. Or, in some tellings of the story, Eos did. In any event, one of them asked Zeus to grant Tithonus immortality so they could be together forever.

The problem, though, was that neither of them thought to ask for eternal *youth*. That proved to be a serious oversight. And Zeus, no doubt chuckling quietly as Greek gods always seemed to do, didn't suggest it either.

I can imagine that deep, throaty laugh echoing down from Mount Olympus. "Immortality? That's all? You got it. No problem."

And true to Zeus's word, Tithonus lived forever. No problem.

Although Tithonus couldn't die, he nevertheless continued to age. Gradually he became frail, weak, and incoherent. In the end, he became an embarrassment to Eos, and she locked him away:

> When hateful old age had utterly overcome him, and he could not move or lift his limbs, to her this seemed the wisest counsel; she laid him in a chamber, and shut the shining doors, and his voice flows on endlessly, and no strength now is his such as once there was in his limbs.

Lifesaving technology can revive a patient like Joe, at least in the short term. But it can't make him young and healthy. Neither can it cure the other maladies that come with age. So we are left with many questions.

What happens when we test the boundaries of life? Do we get a miracle? Or is Zeus still up there somewhere, wondering why we humans never seem to learn?

Should we try to bring people like Joe or Michelle back to life just because there's a very small chance that we might succeed? Will they thank us if we do?

Why does resurrection seem so easy? And how have we convinced ourselves that efforts to resuscitate someone should be automatic?

What does the future hold? And should we be excited or frightened? Or maybe both?

The answers to these questions aren't easy, or simple. So don't think of this as a self-help book. In fact, if you found it accidentally shelved in the self-help section, in between titles on sex and spirituality, just walk away. Now. The same advice holds if you found it filed under "Home Repair: Electrical."

It's precisely because these questions are difficult that the pursuit of answers has led me on a fascinating journey.

I'll tell you about a device that saves people's brains by cooling their noses, and about a bed that supposedly re-creates the heart's natural circulation by shaking people like the Magic Fingers massage beds of the '70s. I'll also show you devices that will shock your heart back to life, and a drug that just might put people in a state of suspended animation.

I'll introduce you to the people behind these inventions. We'll meet a scientist who is convinced that lemurs—small, furry primates about the size of a plush dog toy—can teach humans to hibernate. We'll meet a man who chops off people's heads and freezes them, with a plan to thaw them in a thousand years or so. And I'll introduce you to the drummer in a Journey tribute band who has made a career out of teaching CPR to kids.

We'll also meet a few people who have benefited from the science of resuscitation. There's a man who died a dozen times and who shocked himself back to life, a two-year-old girl who was brought back to life by her nine-year-old brother, and a mushroom farmer who survived a harrowing operation because his brain was put on ice.

We'll also look at some of the hazards to this science. For instance, I'll tell you about a woman whose "natural" death, without CPR, caused a national uproar. I'll introduce you to paramedics who worry every day about whether bringing people back from the dead is the right thing to do. And we'll meet a woman who blames herself because she didn't do enough to save her husband's life.

Although the science that makes resuscitation possible is amazing, its costs—financial, ethical, and emotional—can be enormous. As new breakthroughs continue to emerge, each of which will be more spectacular than the last, those costs are going to grow as well. So we'll all need to think very, very carefully about when, and how, this science should be used to restore a life.

I can't tell you whether you should embrace the science of resuscitation, and I can't tell you whether you should let it bring you, or a family member, back to life. I can only suggest questions that we should all be asking about these advances that are blurring the line between life and death. For now, I can promise you a fascinating glimpse of what the future holds, and the sorts of choices we're all going to need to make someday.

2

Why Amsterdam Used to Be a Good Place to Commit Suicide

Two guys and a dead woman walk into a bar . . .

The year is 1769. April 17, at 9:30 in the morning, to be precise. The place is Amsterdam. And Anne Wortman is lying facedown in the otherwise picturesque canal. She is quite dead.

Passersby stare. A few remove their hats out of respect. More than a few turn away.

Then Andrew de Raad and Jacob Toonbergen emerge from a side street and shoulder their way through the crowd. Respectably dressed in wool frock coats, waistcoats, and breeches, they could be merchants. But what are they doing here?

These two good citizens of Amsterdam tumble down the slippery stone steps and grasp Ms. Wortman roughly by her arms and feet. They drag her up and over the canal's edge, laying her down on the cobblestones. Then they look carefully for signs of life.

Breathing? Nope.

Heartbeat? Afraid not.

Movement of limbs? Opening of eyes? Twitching of fingers? All negative.

Alas, it is becoming increasingly clear that our brave duo has encountered someone who has succeeded in expiring.

Given this information, most Good Samaritans in the eighteenth century—and I'm guessing that's upward of 99.9 percent—would simply have walked away. They'd shrug, check their pocket watches, and remember they have tickets to Christoph Gluck's new opera, *Alceste.*

But not these Good Samaritans. Oh, no.

These men view the victim's departure from the world of the living not as the end of their efforts but rather as the beginning. This, they're thinking, is a challenge.

First, they drape her over a log and roll her back and forth for about fifteen minutes in an attempt to remove some water from her lungs. As they hoped, water emerges. But Anne Wortman fails to wake up.

So the two impromptu paramedics hoist the dead woman. The crowd, perhaps sensing that things are about to get interesting, lends a hand. Soon the dead woman and her rescuers are surrounded by a scrum of bystanders who, for reasons that are not immediately clear, all seem to want to help transport a corpse.

Minutes later the throng arrives at a nearby tavern. The late Ms. Wortman is carried up the stairs, where she is deposited on a bed.

Then the apothecary arrives. Mr. Bernard Donfalaar, it seems, is an old hand at this resuscitation business, and he immediately takes charge. Poor Ms. Wortman is "cold and stiff, without either respiration or pulse. . . ." Admittedly it's not an auspicious beginning. Yet Donfalaar is undeterred.

Under his direction, the inn's servants undress Ms. Wortman and place her before the fire, between two blankets. Next, Mr. Donfalaar begins to rub Ms. Wortman's body with ammonia and spirit of rosemary. He does so assiduously, paying particular attention to "the spine of the back, muscles of the neck, the whole head and temples, and particularly the loins." He also puts a bottle of boiling water on her feet.

Under his direction, Mr. Donfalaar's assistants remove a knife from its sheath and cut off the sheath's tip, creating a flattened hollow tube. They then use this jerry-rigged device to blow tobacco smoke into Ms. Wortman's rectum. One can only assume that Ms. Wortman herself is still a corpse at this point, and thus is more or less indifferent to the liberties being taken with her person.

Somewhere in this flurry of activity, Mr. Donfalaar takes a step that is even more unusual, and which no doubt causes bystanders to gasp in astonishment. He places a handkerchief over the mouth of the dead woman, bends over, and exhales. There is, alas, no immediate salutary effect from this maneuver. Nevertheless, he repeats this process several times. The crowd is breathless.

Unfortunately, so is Ms. Wortman. Things are looking very grim indeed, and the good Mr. Donfalaar's reputation appears to be hanging in the balance.

Finally, perhaps despairing of the possibility of returning this dead woman to the bosom of her family, Mr. Donfalaar brandishes his pièce de résistance. This is the device that will haul her back among the living. This is the newest technology in the new science of resuscitation.

Mr. Donfalaar brandishes . . . a feather.

You can just hear the knowing murmurs circulating in the crowded tavern.

"Told you so."

"Yup, knew they'd have to pull out the feather eventually."

Mr. Donfalaar opens the dead woman's mouth and inserts the feather. Farther. Farther. As far back as he can get. Then he tickles the back of the dead woman's throat.

This is when Ms. Wortman seems to decide that enough is enough. Willing, perhaps, to endure a wide range of indignities, this was beyond anything she'd bargained for. And so, overwhelmed by these ministrations and the kind intentions that inspired them, Ms. Wortman decides that this is an excellent time to rejoin the living.

Admittedly, she does this with something less than perfect grace, choking and coughing and vomiting up copious amounts of brackish canal water. Her rescuers, delighted by this turn of events, ply her with Jenever, a Dutch version of gin made with malt liquor. She responds to this overture—not surprisingly—with more choking, coughing, and heaving.

Her behaviors, everyone realizes, are not usually observed in a dead person. This realization is cause for rejoicing. And rejoice they do.

One advantage of carrying out resuscitations in a tavern is that the ingredients for a celebratory party are mercifully close at hand. Conversely, if things don't turn out well, solace is nearby as well. Either way, those pragmatic Dutch seemed to have had their bases covered.

Now, though, the recently dead Ms. Wortman probably begins to wonder why she is in a strange bed. The fact that she is as naked as a jaybird has probably also begun to penetrate her consciousness, perhaps providing some cause for uneasiness. If she is paying attention, the crowd of onlookers surrounding her has probably also been brought to her attention, further confusing matters in her already-addled, formerly deceased mind. Given all of this, I would not be at all surprised if she begins to at least entertain the idea of returning to being dead.

But she does not. In fact, Ms. Wortman is none the worse for wear. "By these prudent methods," we're assured, "[Donfalaar] had the satisfaction to see her entirely recovered in less than a fortnight." Entirely recovered, that is, except for a newly acquired fear of feathers that she will find herself unable to explain.

Welcome to the strange new science of resuscitation.

THE AMSTERDAM SOCIETY IN FAVOUR OF DROWNED PERSONS

The small stone bridge that connects the two sides of Egelantiersgracht like the horizontal bar of the letter *H* is at the picturesque heart of the most picturesque Jordaan neighborhood of the already quite picturesque city of Amsterdam.

It's a cloudless spring morning and I'm standing by the bridge's railing, surrounded by a veritable United Nations of tourists photographically recording the scenery from every conceivable angle. One couple has focused on a bright red door whose lintel is wreathed with a manicured arch of white roses, and a serious-looking woman in a Yankees baseball cap is in digital rapture over the basketful of tulips attached to an old sturdy bicycle tethered to the canal railing. A crowd of Italian kids feverishly snaps pictures of one another.

I'm painfully aware that I seem to be the only tourist in this beautiful city who is not amassing a digital record of everything in view. Instead, what I'm looking for here in Amsterdam is something that isn't visible in the shops and cafés and tall, lean, row homes that line the canals. What I'm looking for is a bit of history.

I approach the woman in the Yankees cap and mention casually that we're standing in the city where the first successful attempts at cardiopulmonary resuscitation took place.

Her expression suggests that she wasn't aware of this fact. Her expression also suggests that this information isn't rocking her world.

Undeterred, I press on.

Back in 1767, two years before the Anne Wortman incident, a band of concerned Dutch citizens formed the Society in Favour of Drowned Persons here in Amsterdam. Most of the biggest advances in medicine in the past hundred years—anesthesia, surgery, artificial hearts, and of course resuscitation—started right here, I tell her.

But it seems she isn't impressed. With a final look at her beloved bicycle, she's gone. I wander on too, through the sparse traffic to the north side of Egelantiersgracht, then head east toward the center of town. Crossing canal after canal, and looking down into the murky water that seethes between bridge pilings, I begin to get a new respect for those daring souls in the Society who decided that a body floating down there amid the trash might be brought back to life.

Unfortunately, there's little written history about that group or its founders. All we know is that at some point in the 1760s, a small group of merchants, government clerks, and clergy became concerned about the number of their fellow citizens who were drowning in the city's canals.

They met, they debated, and they made lists. Most important, they tried to figure out how best to revive someone who was "apparently dead." Along the way, they kept copious notes of rescues like that of poor Ms. Wortman. We know about her story, and the stories of dozens of people like her, because a British physician named Thomas Cogan who went to medical school in Holland took the time to translate the Society's notes into English.

Some of the rescues he describes were successful. Some others not so much. And yet those feisty Dutch persevered. In their mission to revive the dead, they undertook various experiments. They discussed, they learned, and they tried again. And again. And again.

The Society made strident recommendations regarding effectiveness. And they disseminated those recommendations widely in the form of pamphlets distributed to "hospitals and public charities, coffee houses, taverns, &c." Here we see formal endorsements of many of the techniques that were inflicted on Ms. Wortman.

For instance, inflating the intestines was felt to be a good thing to do. Specifically: "To blow into the intestines through a tobacco-pipe, a pair of bellows, or the sheath of a knife, cutting off the lower point." Time is of the essence. "The sooner this operation is performed with force and assiduity," the Society's pamphlet says, "the more useful it will prove." Better still, the Society suggests that rather than inflating the intestines with mere air, try "the warm irritating fumes of tobacco."

The same document offers other suggestions, including the use of "strong frictions all over the body, particularly down the spine of the back of the neck to the rump." Spirits of ammonia were popular too. Sprinkling a victim with salt was thought to be salutary, as was rubbing with a cloth steeped in brandy. One has to wonder at the culinary spectacle this creates—pulling a dead person out of the water, tenderizing, marinating, and sprinkling him with salt.

The foregoing list might seem to indicate that the Society had a naive credulity regarding any treatments that might conceivably bring a dead person back to life. I mean, if smoke blown into the rectum makes the cut, then the selection process can't be very stringent. Was there anything that the Society *didn't* consider worth a try?

Indeed there was. It was particularly critical of the technique of rolling a victim over a barrel or log, one of the many ministrations to which our young Ms. Wortman was subjected. The Society points out repeatedly in its meeting notes that this method could lead to internal injury and death. Its case studies are full of gruesome accounts of victims who, after being so rolled, proceeded to void blood "from above and below" until they died. Really died.

Alas, though, this rolling technique had so thoroughly worked its way into the minds of the populace of Amsterdam that the Society had difficulty curbing its use. "Yet in vain is it condemned by persons better instructed," the Society laments. "The populace will not easily renounce prejudices supported by long custom." Despite its best efforts, the Society found it impossible to prevent well-meaning rescuers from reaching for a barrel at the first sight of a dead person.

On the other hand, in what has to be its finest moment, the Society seems to have hit a home run. It also recommended an odd procedure that is not formally named. In fact, it is not described in much detail at all, but you'll get the gist: "Let one of the assistants, applying his mouth to that of the drowned, closing the nostrils with one hand, and pressing the left breast with the other, blow with force, and endeavor to inflate the lungs." So: mouth-to-mouth resuscitation was promoted all the way back in the eighteenth century.

And the Society was bullish about its benefits. "We believe," the manual opines, "that, from the beginning, this might be as efficacious as blowing up the fundament [rectum]." Indeed. And much more pleasant, one can only assume, for everyone involved.

The Society even offered a suggestion that the American Red Cross has, alas, not taken up. With typical Dutch caution, the Society advised placing a handkerchief over the victim's mouth. When placing your lips onto lips that have been immersed in a canal for who knows how long, and when those lips are connected to a body that is quite possibly not just "apparently dead," but really dead, I'd say a handkerchief—at least—is a very good idea.

Finally, if all else fails, you could tickle the back of the victim's throat with a feather. This is my own personal favorite, the technique that onlookers might have said did the trick in the case of Ms. Wortman, but I deeply regret that the science behind it has since proven to be nonexistent, as we'll see.

Actually, to be fair, there is at least a bit of hard science behind most of the Society's recommendations. Some methods, in fact, have been repurposed and used in different ways more than a century and a half later. But more about that later.

The Society was also methodical in measuring its outcomes. In the first two years of the Society's existence, in fact, it claimed to have saved more than 150 people who were "apparently dead." Unfortunately, we'll never know how many of those "victims" were really dead, or even apparently dead. One can imagine that a society eager to demonstrate its impact might be tempted to count a few questionable cases as victories. So during those heady first years when members of the Society were on the hunt for victims to rescue, it probably wasn't a good idea to take a nap in public, unless you wanted to wake up with a very uncomfortable tickle in your throat.

The most important contribution of the Society, though, was arguably its exhortation to leave no dead body unturned. "Since no body can affirm with certainty that the drowned be really dead, unless there be some signs of putrefaction on the body, we hope that till then all possible efforts may be made; and that those who know any other means of assistance will communicate them to us."

Admittedly, that admonition sets the bar pretty high. Unless a body is rotting, these guys suggest, it's worth setting it in front of a fire, tickling it, and breathing into its orifices. Which is fine, I guess, if you happen to be that body. But it was probably a rough time to be a Good Samaritan.

LATEAT SCINTILLULA FORSAN—"PERCHANCE A SPARK MAY BE CONCEALED"

Around the same time that the Amsterdam Society was advancing the science of resuscitation, Londoners, too, were becoming increasingly concerned about their fellow citizens' fatal encounters with various bodies of water in and around their fair city. One popular place for drownings and near-drownings was Hyde Park's Serpentine, the long lake that slashes diagonally across the park. So in an effort to save some lives—and perhaps motivated by a friendly sense of competition with the Dutch—Londoners built a "receiving house" in 1794. Designed as a sort of emergency room and research laboratory, this structure was the stage for hundreds of miraculous resuscitations, and was the proving ground for new therapies.

After a morning spent in the dusty London Public Library, I have in hand a copy of an article from the *Illustrated London News* that describes this building as "a neat structure, of fine brick fronted and finished with Bath and Portland stone." So that's what I'm looking for as I stroll through Temple Gate into the heart of Hyde Park. That "neat structure" has to be here somewhere.

Although I have a description of the receiving house's exterior, it's what's inside the building that is truly unique. According to the article, the Royal Humane Society—a knockoff of the Dutch society—spared no expense in building it. It has an entrance hall and separate wards for male and female patients. And, luxury of luxuries, the *Illustrated London News* boasts that this place has "beds warmed with hot water, a bath, and a hot-water, metal-topped table for heating flannels, bricks, &c." One is left to imagine what the "&c" might be. It is, the article concludes with rather un-British immodesty, "a model for all other institutions of the same kind." I can hardly wait.

Just past the gate, there's nothing in sight that looks like it might be what I'm looking for. But it's a cloudless Sunday afternoon, and the weather is bringing the local inhabitants out in mobs for the first glimpse of sun in what perhaps has been years, judging from the rapt, pale faces that ghost past me.

As I join the sunstruck Londoners, it's easy to see why the Serpentine became a hot spot for the resuscitation movement. Created in 1730, this twenty-eight-acre body of water sits right in the center of London. That makes it an unavoidable water hazard for strollers, boaters, cyclists, and—in the winter—ice-skaters. In fact, shortly after its creation, Londoners flocked to the Serpentine to see and be seen and, with astonishing frequency, become apparently dead. That, it turned out, was the motivating force that catapulted the quiet science of resuscitation into the public view.

The story of the Royal Humane Society begins in 1773 with a London physician, William Hawes, who became curious about the techniques used to bring the apparently dead back to life. At first, it seems that he hadn't heard about the work of the Amsterdam Society. Whether he was entirely ignorant of the Dutch techniques is unclear, but when you're dealing with a dead body, the list of plausible revivification maneuvers is rather short. So perhaps it shouldn't be surprising that Hawes entertained many of the same ideas that the Dutch had tried: mouth-to-mouth resuscitation, warming, and tobacco featured prominently in Hawes's repertoire as well.

But Hawes didn't want to merely try these interventions. He was also curious. He wanted to figure out whether they worked.

Hawes was so curious, in fact, that he wasn't content to limit his experiments to the victims who happened to show up on his doorstep.

So to supplement these meager and unpredictable offerings from the deep, he hit on the somewhat ghoulish expedient of paying anyone who would bring him a body rescued from the water "within a reasonable time of immersion." Think of it as the human version of a deposit on bottles. (The Dutch did the same thing, but being of a practical mind and penurious disposition, they only paid for successful resuscitations.)

History, alas, is silent on what constituted "a reasonable time." One can imagine that the offer of payment might have induced more than a few cash-strapped entrepreneurs to stretch that definition just a bit. But Hawes got his bodies, and more than a few people earned a modest living. Someone might even have survived as a result of Hawes's ministrations.

A year later, in 1774, Hawes met up with Thomas Cogan, the physician who had translated the results of the Amsterdam Society's work. What was probably a very morbid conversation ensued, and on April 18, 1774, the two each brought fifteen citizens—colleagues, church officials, merchants, and minor royalty—to the Chapter Coffee House in St. Paul's Churchyard, where they founded what was to become the Royal Humane Society.

The group had prestige, social standing, and money. It also had a great motto: "Lateat Scintillula Forsan." In case your Latin is as rusty as mine is, this is often translated as "Perchance a Spark May Be Concealed." The motto is supposedly carved over the entrance to the receiving house.

Speaking of which, now I've passed the Ring Bridge and made my way along what is called the Long Water to the far northwest corner of the Serpentine. I've been dodging bladers, cyclists, and matrons pushing prams, without seeing a hint of the neat structure of the receiving house. Still, I tell myself, it's a beautiful day. I keep walking.

In Hawes and Cogan, the Society also had a pair of brilliant marketing strategists. First, they'd done their homework and had calculated that the previous year in London alone, 123 people were known to have drowned. If they could save even half, they said, or even one-tenth, that would be a spectacular victory. In appealing to the public for funding, the Society used this number freely, but it also made its appeal personal. For instance, it asked for support in restoring "a father to the fatherless, a husband to the widow, and a living child to the bosom of its mournful parents." Not surprisingly, the money rolled in.

What followed was, among other things, the development of an elaborate scheme that the Society concocted for bringing in bodies and for "receiving" a body in a tavern, which would become a theater for resuscitation. It was a living for a few and a scam for many. The going rate for what the Society paid out, by the way, was two guineas (about $250 in today's money) to a bystander who attempted resuscitation, four guineas when that attempt was successful, and one guinea to an innkeeper who offered up his establishment for a resuscitation party.

Eventually, the Society stopped paying for dead bodies dumped on its doorstep. History is not forthcoming about when, exactly, this payment system ended. Nor is there a clear explanation of why, although one can imagine.

Instead, the Society began to spend its charitable contributions on innovation and on the testing of new "scientific" forms of resuscitation. For instance, it invested in a bewildering variety of new bellows for inflating the lungs. It also seemed to be particularly enthusiastic about the potential benefits of various heated beds and chairs that could be used to warm bodies quickly. But the pinnacle of its investment was the "receiving house."

It was built not merely to save the lives of unlucky bathers in the Serpentine, but also to ensure that new methods of resuscitation could be tested under controlled circumstances (unlike the circumstances found in inns and taverns).

The Society's seventieth annual report published in 1844 boasts that a wide range of promising techniques received "a fair trial at the

Receiving-House in the Park." Only the most efficacious, it claimed, were selected. Of these, some of the strongest recommendations are reserved for warming the body, and the application of friction with "rough cloth or flannel." The use of bellows with a volume of approximately 40 cubic inches is also recommended, but only by "Medical Men, who may all be fairly presumed to be perfectly conversant with the modes of carrying on artificial respiration."

I'm not so sure this was a fair assumption to make, but the language of science is truly impressive. And reassuring. The early nineteenth century, it seems, was an excellent time to drown.

In fact, reading the Society's annual reports that describe the results of these experiments gives a remarkable sense of scientific advancement and rapid discovery. For instance, its seventieth annual report (1844) describes a new revelation in the science of artificial breathing. "For some time," it says, "the practice has prevailed of inflating the lungs by means of a curved tube passed through the mouth into the trachea."

Of course, you think. Who would dream of inflating the lungs in any other way? Well, you're in for a surprise.

"It is now deemed more advisable," the report proposes delicately, "to introduce the air through a short ivory pipe inserted into one of the nostrils, pressing back the lower part of the larynx upon the commencement of the oesophagus, to prevent the air from passing down into the stomach." Who would have thought? That section of the report concludes matter-of-factly: "The metallic trachea tube is, therefore, discontinued, and the instrument-case is furnished with ivory nostril pipes."

So there you have it. No more curved tubes. We'll all use ivory nostril pipes.

And there's some solid reasoning to support this. Even now, when inserting a breathing tube is difficult (for example, in a patient who is overweight), a nasal tube is often much easier. It's also a more reliable technique when the person trying to inflate a victim's lungs is inexperienced, as most doctors in the eighteenth and nineteenth centuries no doubt were.

However, the Society produced other recommendations that didn't make much sense. For instance, an early pamphlet tells would-be rescuers: "Grasp the patient's arms just above the elbows, and draw the arms gently and steadily upwards, until they meet above the head." This, the pamphlet explains helpfully, "is for the purpose of drawing air into the lungs." Next, the pamphlet tells rescuers to "turn down the patient's arms, and press them gently and firmly for two seconds against the sides of the chest. (This is with the object of pressing air out of the lungs.)" Finally, just in case you're tempted to walk away, the pamphlet reminds you that you're not done yet: "Repeat these measures alternately, deliberately, and perseveringly, fifteen times in a minute." Although I'm certain that these maneuvers put on quite a show for bystanders, I regret to report that they probably did no good whatsoever.

In fact, while it was most active during the nineteenth century, the receiving house was host to trials of dozens of resuscitation techniques, many of which were no more effective than flapping a person's arms. (Rescuers continued to protect bathers and skaters well into the twentieth century. Victims were hung upside down, for instance, and rolled over barrels. The Amsterdam Society's reservations about this latter approach apparently didn't dissuade the British from trying it, although they, too, would abandon it eventually.) Other strategies involved being thrown over the back of a horse and trotted at a brisk pace and, of course, tickling the back of the throat with a feather.

The use of these techniques may make it seem like the work of the receiving house was little more than charlatanism, but thanks to the Society's meticulous record-keeping we know that it was able to demonstrate remarkable successes by the end of the nineteenth century. For instance, in 1884, of 270,000 bathers in the peak bathing months of June and July, there were thirty-one rescues. Of these, fifteen were taken to the receiving house, where they were successfully brought back to life from apparent death. Even today, those statistics would be impressive.

All of these advances took place in the receiving house that I seem to be unable to locate. So finally, I decide to ask for directions. I pick my target carefully—an older man in his seventies dressed in gray flannel from head to toe. He's sitting on a secluded bench with a commanding view of the bridge and the length of the Long Water, and he's resting a thin, fine-veined hand on a polished mahogany cane. I'm thinking that he doesn't look too busy.

He considers my question for just a second, and then his eyes light up under overgrown eyebrows. He nods.

"That was brilliant, that was. They were always rescuing someone. Racing around, pulling people out of the Serpentine and bringing them back to life."

I ask if he actually saw them. Was he there?

"I certainly was." He nods. "My father would bring us here when we were lads, back in the '30s. We learned how to swim just over there." He gestures to his right, along the southern bank, where the Lido stands now. In the eighteenth century, that had been exactly where the park ranger, the Duke of Cambridge, decreed that swimming should take place. The bottom there had a gentler slope, he argued. And by corralling bathers into a small area, he suggested, the lifeguards would be able to watch them more closely for signs of trouble.

"Those were the days."

I ask him if he ever saw anyone drown, but no sooner have the words left my mouth than I realize that this sort of question suggests a certain morbid fascination, and thus is perhaps not the best one with which to pursue a conversation with a perfect stranger. But the gentleman in flannel seems unruffled, and even a little excited.

"Oh my, yes. All the time. That was half the fun of coming here. You'd always be keeping an eye out for someone who might be in trouble.

Then the lifeguards would come rushing in. Or the ice men, in the winter. They'd pull the gent right out. Other times, they'd lay him over the back of a horse and trot him up to the house."

Ah. The house. That could only be the receiving house. At last.

I feel like an explorer who finds the last survivor who can point the way to a forgotten Mayan temple. Here was someone who had actually witnessed these events. And someone who, apparently, was as titillated by them as I was. Even better, someone who knew about the receiving house.

But first, I have to ask whether he ever saw them revive anyone . . . with a feather?

The man looks at me blankly for a second, but then his face breaks into a grin revealing two rows of perfectly immortal dentures. "Well, no, I can't say that I did. But I heard about a whole list of other things." He pauses, lost in a reverie. "Let's see. There was the trotting on a horse, as I said. That was great fun. Helped get water out of the lungs, or some such. And there were bellows—all manner of bellows that were supposed to breathe for you. And then, sometimes I heard that they'd roll people over a barrel—back and forth." He slides one elegant hand back and forth over the crook of his cane to demonstrate, then shakes his head. "I'm not sure any of that worked, but it was quite a spectacle."

"And the receiving house? Where they took people? Is it close by?"

He shakes his head. "Oh, no, that came down years ago. In the 1950s, I'd say. Maybe earlier. No need for it, you see. There were fire trucks and ambulances. So they tore it down. Pity." (Actually, I learned later from an authoritative source that it was destroyed in the London Blitz in 1940.)

Pity indeed. Well, so much for that. But at least I've been rewarded with a brush with history, and a pleasant walk. I thank the gentleman and thread my way northward and out of the park.

Wandering out of the relative quiet of Hyde Park and onto the streets of London, I'm thinking about all the techniques that were tried in the receiving house. And those stories were making me curious. Could tobacco smoke in the right place really revive someone? And what about being hung upside down?

Intriguing. But a little scary. And that's a problem, because I'm beginning to get the idea that I should try one of these techniques myself.

If that sounds like an odd way to spend a morning, I suppose it is. But these stories of resuscitation seem so distant—and many of them so bizarre—that I want to take a firsthand look at how they work. And I really want to understand why anybody thought that they might work at all.

I mentally review the list of the receiving house's recommended treatments, looking for something that isn't illegal, stupid, or downright dangerous. Nope. Nothing.

So then I go back over the list looking for something that at least won't kill me, and that isn't likely to be too uncomfortable. Now, that's a little better. By the time I pass through Speakers' Corner and down into the Underground, I have it whittled down to two possibilities.

I could suffer a near-drowning accident and then have someone waiting nearby who is ready to tickle the back of my throat with a feather. That might be fun. I have fleeting visions of being carried from the surf by a naked Amazon, who would whip out a feather—from somewhere—and . . .

But some fantasies are best left on the shelf. Besides, there's the whole "near-drowning" thing. That doesn't sound very appealing. Next.

I could fling myself across a horse's back to see if that's a reasonable substitute for breathing. That doesn't take much deliberation. A mild morning on horseback seems like the easy way out.

My choice is made a little easier by the fact that the science behind this particular maneuver is plausible. In theory, at least, the trotting horse would provide the same sort of effect on a dead person's chest that the compressions of CPR produce. In addition, that same up-and-down bouncing might move the chest wall enough to cause air to flow into and out of the lungs. But would it work?

HORSEBACK-RIDING LESSONS

A week later, I find myself looking up at a very large horse named Penny, and down at a compact, wiry horse trainer whom I've been instructed to call "D." D doesn't seem to talk, and he doesn't smile. He doesn't even have much facial expression at all. But on the bright side, he doesn't seem to be fazed by my request to be strapped across the back of a trotting horse.

It took me three failed attempts before I realized that calling a riding stable and asking point-blank to be strapped to the back of a trotting horse was bound to end in disappointment. Finally, I got an appointment by asking if they offered riding lessons, without being very specific about exactly what I had in mind. The nice lady on the phone agreed, not entirely aware of what she was agreeing to. Alas, neither was I.

So there I am, standing in the middle of a round, dirt-floor riding arena that's about 100 feet in diameter. There's a sturdy wooden partition around the edge to prevent horses from escaping, which offers me some degree of reassurance. There are also half a dozen horsey people watching us surreptitiously, which doesn't. Out of the corner of my eye, I think I see a child point at me and laugh. I begin to suspect maybe this isn't such a good idea.

Actually, I'm pretty sure this isn't a good idea. In the past week, I'd e-mailed several colleagues, hoping to get some sense of what this exercise might entail. They weren't encouraging. Their answers ranged from pointed questions about my maturity ("Does your family know you're doing this?") to dire warnings about potential side effects, including a nosebleed, heartburn, nausea, cardiac arrhythmia, and respiratory arrest.

I'm hoping I can get away with just a nosebleed. That would be harmless but impressive. I'm decidedly less enthusiastic about the other possibilities.

D points to my left foot, and then to Penny's left stirrup. Then he points to my left foot again, just in case it wasn't quite clear what's expected of me. Carefully navigating around a pile of steaming manure that lies in front of Penny's left rear hoof, I do as I'm told. A moment later I find myself draped over Penny's back, my cheek flattened against a dirty blanket that has the texture of greasy steel wool. D is tying my feet to my wrists under Penny's belly. I'm guessing he's improvising, but he seems confident.

The world, meanwhile, has turned upside down. Then D gives Penny's rump a hearty slap. Then the upside-down world starts to move.

Slowly at first, in lurching bumps. Then Penny's pace quickens and we're trotting merrily around the paddock. My head is on the inside of the circle and my feet are on the outside, which means I've lost sight of D and everything else, for that matter. All I can see is Penny's muscular right haunch, swinging back and forth like a pendulum. A very energetic pendulum.

The sensation of being trotted around a riding arena isn't entirely unpleasant at first. There's even something rhythmically hypnotic about being bounced up and down, and up and down. In fact, it's relaxing me enough that my sense of time is getting hazy. I can't tell whether I've been trotting for a minute or five, although some part of my brain is still making a mental note whenever we pass the pile of horse manure I almost stepped in when Penny and I were first introduced. That landmark has appeared two times so far, which means two circuits around the ring.

At some invisible signal from D, Penny's rhythm shifts into a higher gear, and the ride is no longer a lulling bounce. It is, I'm thinking, more like how you might feel if a basketball is being dribbled on your stomach. By a very large man with hubcap-size hands. Who has an anger management problem.

This is my first thought.

My second thought is that it's becoming difficult to breathe. Very difficult. As in: "Help! Help! Oh my God, I'm going to die!" That kind of difficult.

After a few panicked moments of hyperventilation that do no good whatsoever, I stop trying to breathe. I just stop. And I even relax a bit. Hypoxia, I'm told, has that effect on people. At least, it does on me.

But then something interesting happens.

When I was trying to breathe on my own, I was fighting Penny. And Penny, needless to say, was winning. But as soon as I relax, and stop trying to breathe at all, I've discovered that Penny—bless her heart—is breathing for me. In and out, in little, panting dog breaths. Big enough breaths, it seems, to move some oxygen. Big enough breaths, apparently, to keep me alive.

But how does that work? The mechanics of breathing are really pretty simple, and it wouldn't be difficult to replicate them with a horse. We breathe mostly with our diaphragm—a web of muscle that stretches across the lower edge of the rib cage, separating the chest from the abdomen. When the diaphragm contracts, it creates negative pressure (a vacuum) in the chest cavity. That drop in pressure causes the lung's alveoli (small air sacs in the lungs) to pop open, pulling air in through the trachea. When the diaphragm relaxes, the pressure increases and air flows back out.

A horse can't reproduce the diaphragm's natural movement, but it can move the diaphragm and chest wall in a similar way. For instance, as the horse's back rises, it presses in on the abdomen, forcing air out of the lungs. It has much the same effect on the chest wall, pushing in and further emptying the lungs. Then, in the second or so that an inert body is bounced up and is airborne over the horse's back, the diaphragm and chest wall bounce back to their usual shape, causing air to flow back in.

Indeed, this seems to work. Chalk one up for the Royal Humane Society. The trotting-horse method sounds like it might be a success.

Well, only a partial success. Because taking in oxygen is only half the work of breathing. The other half, getting rid of carbon dioxide, is just as important as the first, because too much carbon dioxide will kill you just as quickly as not enough oxygen will. As carbon dioxide builds up, it turns into carbonic acid in your blood, making your blood acidic.

Although respectable medical textbooks will never describe the acidification process in these terms, you can think of it as somewhat like mainlining Pop Rocks. This, in turn, wreaks havoc on your body. Also, to add insult to injury, high carbon-dioxide levels make you feel short of breath, even if you're getting enough oxygen.

These, anyway, are the little lessons in physiology that are flitting through my fading brain as I remember that to get rid of carbon dioxide, we need to take deep, full—and *slow*—breaths. That's something I most certainly am not doing. Apparently, Penny has not been schooled in the mechanics of respiration and gas exchange, and so it seems that her graceful trotting is going to kill me.

Carbon dioxide was first described by a Scot, James Black, who carried out all sorts of cunning experiments involving weights and balances and balloons that I'm finding very difficult to remember as I continue to be bounced along by Penny. He called carbon dioxide "fixed air." He lived to be seventy-one. He never married but had numerous very close male friends. Odd what facts you remember when you're on the verge of blacking out.

I try waving at D, only to realize that my hands are tied. That this comes as a surprise to me should give you some sense of my cognitive state. It should also be a warning, just in case you still need one, that my little riding experiment is one that you really, really don't want to try.

After what seems like a lifetime, Penny's pace slows, D's boots reappear, and I start breathing for myself again.

As Penny downshifts to a walk, I can't help noticing that she isn't even breathing hard. On the other hand, I'm gasping like a beached fish. Or like someone who was very close to becoming "apparently dead." As we stop, D loops Penny's reins over a post and unties me.

I stand tall and stretch. I pat Penny's flank with what I imagine is a horsey gesture. Then I throw up, violently and copiously, narrowly missing D's boots but splattering my own sneakers beyond redemption.

Penny seems unperturbed, but D takes a step back, impressed. I'm more than a little lightheaded, and mostly just trying to stay upright. But I swear, just for a second, I see him smile.

FEATHERS, WHIPS, AND ICE: THE EARLY SCIENCE OF RESUSCITATION

That was not my finest hour. But at least Penny seemed to have enjoyed her little romp. And I'd certainly brightened D's day.

More important, I learned something. If you're relaxed enough—that is, if you're hypoxic enough to be semiconscious—the trotting motion of a horse might actually breathe for you. At least for a little while. On the other hand, it seems unlikely that fine animals like Penny will ever have a place on a hospital's "Code Blue" team.

But the British Society did have some successes. Remember its report of fifteen lives saved in the summer of 1884? Something in their toolkit must have worked.

What about fumigating someone's orifices with tobacco smoke, for instance? Those methods, at least, are straightforward. As pioneered by the Dutch, a rescuer lights up a pipe and blows smoke directly into the victim's mouth, nostrils, or rectum.

If we set aside that last option—and please let's do—there's a certain appeal to arriving at a scene of crisis, only to pause, remove a briar wood pipe from one's waistcoat pocket, and embark on the little rituals of filling, tamping, lighting, and puffing. That sort of routine would surely have a calming effect on panicked bystanders and family members, which should be reason enough to try it. But did it work?

It might have. Nicotine, the ingredient of tobacco with the most prominent cardiovascular effects, belongs to a class of chemicals known as alkaloids (of which cocaine is one) that come from the nightshade family. It's also absorbed through mucous membranes in the mouth (or rectum). So far, so good.

The most important thing to know about nicotine, though, is that it's very *rapidly* absorbed. Usually nicotine exists in a non-ionized form that crosses mucous membranes into the bloodstream very quickly. It also crosses from the bloodstream into the brain with a similar rapidity. The result is that smoke blown into someone's mouth will reach the heart and brain in a matter of seconds, presuming there is enough heart function to keep blood flowing. (That's not the case if the nicotine is in an acidic environment, like the stomach, from which it is not well absorbed.)

Nicotinic receptors exist throughout the brain and in the peripheral nervous system, but nicotine has a much higher affinity for brain receptors. That means that although the nicotine you inhale in a cigarette could act anywhere in the body, its effects are mostly felt in the brain. The one exception is that nicotine binds avidly to neurons in the sympathetic nervous system that activate the adrenal glands, releasing epinephrine. Epinephrine causes an increase in heart rate and respiration, as well as an increase in the strength of the heart's contractions. Not surprisingly, injectable epinephrine is one of the key items on resuscitation carts.

So tobacco smoke is certainly a plausible way to revive someone. And that's particularly true in the setting of drowning, because a series of interlocking reflexes effectively dampens the heart's activity. Cold water applied to the face, for instance, causes a drop in heart rate—the so-called diving reflex. Indeed, some people pulled from particularly cold water have a heart rate of only a few beats per minute (a normal rate is from 60 to 100 beats per minute). In those circumstances, a jolt from the sympathetic nervous system is exactly what the heart needs. So it's plausible that tobacco smoke might have been at least partially effective. It probably wouldn't have restarted a heart that had stopped completely. But if a drowning victim had a very slow and weak pulse—perhaps slow enough to be undetectable to a bystander—tobacco smoke might have increased the heart's activity to give the appearance of resuscitation.

Sadly, enthusiasm for blowing smoke into various orifices declined sharply in 1811 because of a killjoy named Sir Benjamin Brodie. Brodie was an English physiologist and surgeon who doubtless had numerous positive traits, but he almost certainly was not an animal lover. In a series of experiments on dogs and cats, Brodie claimed to have determined that tobacco smoke is potentially lethal, and he even identified the lethal dose. (Just in case you're wondering, it's four ounces for dogs, and one ounce for cats.) Despite the fact that no one was seriously suggesting that tobacco smoke be used to resuscitate family pets, people generalized Brodie's findings to humans and decided that the whole smoke-blowing strategy was probably best avoided entirely. Such was the state of evidence at that point in history when shrill activism by someone like Brodie was enough to change clinical practice. In the absence of randomized controlled trials showing that tobacco smoke was safe and effective, its proponents found it easier to cave in to popular opinion.

That's too bad, because as we proceed down the list of potential resuscitation techniques, evidence of effectiveness gets pretty thin. For instance, another widely used method—flagellation—seems to have very little to recommend it. Beating a drowning victim with whips apparently seemed like a good idea to someone, but the historical record is noticeably—and sadly—silent as to whom that someone was. The best that can be said for this approach, I suppose, is that if the person doesn't wake up, well, no harm done.

Another technique advocated by early resuscitation enthusiasts was that trick of tickling the victim's throat with a feather. Unlike with flagellation, where if the victim actually revives during the beating he or she will be quite sore or perhaps have a broken bone or two, the feather technique is likely to do more harm than good. Activating the gag reflex when someone is unconscious, or semiconscious, can lead to vomiting and subsequent inhalation of stomach contents. This unfortunate series of events is known in medical circles as aspiration pneumonitis, which is often rapidly fatal.

If that doesn't kill you, there's a good chance that activating the gag reflex will stimulate the vagus nerve, which reduces the heart rate. Actually, the gag reflex has much the same effect on heart rate and respiration that drowning does, and thus is likely to make a bad situation worse. So feathers, sadly, are out.

It's at this point that you have to wonder—if the British Society took such careful notes, how could some of its recommendations be so implausible? In part, the problem was that many of these resuscitations were reported by people with little or no medical training. So the techniques that were used may have varied widely. More important, without the sorts of tools we have today—an electrocardiogram, a heart-rate monitor, even a stethoscope—it's difficult to tell for sure what these interventions actually accomplished. If a drowning victim like Anne Wortman, who was tickled with a feather, woke up, did that feather restart her heart? Or was she simply unconscious, and did the gag reflex wake her up? The British Society didn't know, and so it probably counted many cases of unconsciousness as examples of successful resuscitation.

That lack of good evidence explains the presence of two common—and contradictory—views of the role that temperature plays in a successful resuscitation. On one hand, the Royal Humane Society was strident in expressing its opinion that the apparently dead should be warmed in the quickest way possible. Immersion in warm water was frequently recommended, as was using blankets, warm sand, or placing the victim next to a fire. The Society—forgetting for a moment its stern Victorian moral code—even recommended the use of volunteers who would climb into bed with the apparently dead. It's not clear whether these volunteers were supposed to be fully clothed, but that's probably best left to the imagination.

On the other hand, at roughly the same time, others proposed what was known, somewhat ominously, as the Russian Method. Rather than putting the victim in bed with his or her fellows, the Russians apparently believed that cold was better. So they would pack victims in ice or cold water. Or they would simply toss them outside.

One has to wonder whether anyone believed that these chilling maneuvers improved a victim's chances. Indeed, it sounds like an ideal ploy to get rid of an enemy, mother-in-law, or czarina. "Don't worry," someone might say, "we'll just put Catherine outside in the snow. She'll feel better in no time. Really."

It turns out that both approaches have something to recommend them. Drowning, for instance, often results in hypothermia—a core body temperature that is 5 or even 10 degrees below the normal 98.6 degrees. Hypothermia is a problem because it reduces the heart rate and respiratory rate. Profound hypothermia, below 93 degrees, also makes the heart's electrical conduction very fragile. It's easily disrupted, and very difficult to convert to a normal rhythm. So there is something to be said for warming a cold drowning victim, since doing so makes it easier to restart the heart. Indeed, there's a saying in emergency medicine that a hypothermic victim who is found unconscious and believed to be dead isn't dead until he's *warm* and dead.

But the Russians weren't entirely wrong, either. It's true, cold will make it more difficult to restart a heart. But as we'll see in chapter 3, hypothermia can protect the brain and other organs by making them less susceptible to damage caused by low levels of oxygen and the subsequent buildup of toxic chemicals and free radicals that can damage tissue.

That debate, in a nutshell, is perhaps the principal conundrum of the science of resuscitation. Colder temperatures make it more difficult to revive a person, increasing the probability that someone who is apparently dead will become truly dead. However, warmer temperatures allow damage to the brain, increasing the likelihood that we'll wake up—if we wake up at all—with the IQ of a Snickers bar.

PREMATURE BURIALS AND "SAFETY COFFINS"

These first forays into the science of resurrection gave us the tools to save lives. But they also gave us another gift. And it's one that we're still living with, 250 years later.

The incredibly common fear of being buried alive is arguably the direct result of advances in the science of resuscitation. Think about it: If someone who is apparently dead can come back to life, what would happen if that person comes back to life a little too late? What if—just for instance—that poor guy comes back to life in a coffin? And if there was one outcome that terrified people more than anything else, it was waking up in a coffin. Underneath a lid that's been firmly nailed down. Six feet underground. The thought alone is enough to give anyone nightmares.

Right now I'm wandering through a cemetery just outside of Williamsport, Pennsylvania, looking for the grave of one person whose nightmares must have been particularly vivid. It's very late on a cold spring afternoon, and the anemic sun has already disappeared behind a wooded hillside. A particularly chilly gust slips through my thin jacket, and I shiver, thinking that an early evening stroll in a graveyard is enough to make a generally well-adjusted person feel a little spooked.

In fact, if I were alone, I would probably turn around and head for the nearest bar in search of a good stiff drink. But fortunately I'm not alone. I'm tramping along behind a tall, gaunt, whip-thin Lincolnesque figure named Gerald, who happens to be a bona fide grave digger. (Here and throughout, I've used first name pseudonyms for people who don't want to be identified.) He is leading me through the sprawling Wildwood Cemetery to the grave of Thomas Pursell, a nineteenth-century fireman whose fear of being buried alive went further than most.

Pursell was so terrified of this fate that he had a five-coffin crypt built for his entire family with felt-lined walls, so if any of them woke up flailing, they wouldn't injure themselves. And he made sure that when he died, which he did in 1837, the vaults would be stocked with bread and water. This was a guy who really planned ahead.

I'm pondering this as I notice, just in time, that Gerald has stopped right in front of me. We're facing a wide stone structure that looks nothing like the tomb of a humble fireman. Across a grand façade of limestone blocks, there are five elaborate body-wide iron doors, with bas-relief figures gracing each one. Copper edging around the doors gives them an added air of elegance. In my opinion the tomb looks more like a fancy pizza oven than a place to keep your dead.

Now that we're here, I can't resist asking Gerald if he was ever tempted to look inside.

He stares at me like I've just suggested breaking into one of his graves, which, I suppose, is more or less exactly what I've done.

I clarify that I'm just asking about looking. I wasn't suggesting . . . grave robbing or anything.

Gerald gives me another couple of moments of uncomfortable scrutiny before he relents.

"Can't."

"Can't?"

"Doors only open from the inside."

I'd like to ask why Pursell would want to keep people out of his tomb. But Gerald is waiting, and so I thank him. He heads back to his unfinished grave, and I try to find my car, still thinking about Pursell's odd obsession.

The oddest thing about Pursell's story, though, is the fact that he was hardly unique. Just as the Royal Humane Society and others were making impressive progress in reviving the "apparently dead," the public was becoming increasingly concerned that they might be buried when they weren't really dead. As William Hawes and his fellow resuscitationists trumpeted their successes in bringing back the dead, the message—clearly and unmistakably—was that if you love your relative, you'd better make good and sure he or she is really, truly dead.

That, at least, is the message that the public heard. All this talk about people being only apparently dead was enough to make a lot of otherwise normal people like Thomas Pursell very anxious. His anxiety even has a name: "taphephobia," or a fear of being buried alive.

One of the earliest symptoms of that growing phobia was a lengthening of the funeral process. Rather than getting a loved one into the ground immediately, in the mid-eighteenth century there was a growing sense that it might be best to wait a day, or two, or longer.

William Tebb, an English businessman, antivivisectionist, vaccination critic, and all-around meddler, took up this cause in 1896, calling for measures to prevent premature burial. To his credit, he brought numbers—data—to the argument. For instance, he produced dozen of examples, culled from letters and word of mouth and newspaper accounts of premature burials and near misses. Consider this one, which Tebb found in *The Undertaker's Journal* (July 22, 1893):

> Charles Walker was supposed to have died suddenly at St. Louis a few days ago, and a burial certificate was obtained in due course from the coroner's office. The body was lying in the coffin, and the relatives took a farewell look at the features, and withdrew as the undertaker's assistants advanced to screw down the lid. One of the undertaker's men noticed, however, that the position of the body in the coffin seemed to have undergone some slight change, and called attention to the fact. Suddenly, without any warning, the "corpse" sat up in the coffin and gazed round the room. A physician was summoned, restoratives were applied, and in half an hour the supposed corpse was in a warm bed, sipping weak brandy and water, and taking a lively interest in the surroundings. Heart-failure had produced a species of syncope resembling death that deceived even experts.

Alas, we don't know exactly what medical condition had convinced "experts" that poor Mr. Walker was dead and gone. It might have been heart failure, or perhaps a profound reduction in heart rate. Or maybe just a really top-notch nap. We'll never know.

But that uncertainty didn't dampen Tebb's anxiety. Indeed, he evidently thought that these sorts of incidents occurred at an epidemic rate: "ALMOST every intelligent and observant person with whom you converse," he said, "if the subject be introduced, has either known or heard of narrow escapes from premature burial within his or her own circle of friends or acquaintances; and it is no exaggeration to say that such cases are numbered by thousands."

Is that an exaggeration? It's difficult to tell. Of course there were stories of circumstances in which it appeared that death was declared a little prematurely. And as we've seen, without diagnostic instruments like an EKG machine or even a stethoscope, it's certainly possible that a few unlucky souls woke from a night out drinking to find themselves in a funeral home. Still, it's safe to say that Tebb's anxious warning about "thousands" of such incidents is probably not realistic.

Nevertheless, it was no surprise that Tebb's will stated unequivocally that he not be put in the ground until his body showed clear evidence of decomposition. Depending on environmental factors like heat and humidity, such evidence can take anywhere from three to seven days to manifest. One can only imagine that anyone making a demand like that, particularly in the days before refrigeration—or air fresheners—was not on good terms with the family he was leaving behind.

Many of these fears were no doubt overblown, but to be fair there are well-documented cases of people who were sent to the graveyard before their time. One of the best known of these is the case of Anne Green, who was sentenced to be hanged for ending the life of her newborn child, born out of wedlock. After the hanging on December 14, 1650, her body was sent to the home of Dr. William Petty, an Oxford instructor in anatomy, who had requested a female cadaver for dissection. Much to Petty's surprise, Ms. Green was not quite dead. Actually, she was very much alive. Alas, we don't know what that moment looked like, but one can imagine that there was considerable consternation on both ends of the dissection knife. Even more surprising was the fact that Ms. Green apparently lived happily ever after. Having been hanged once (one assumes no statutes allowed for a person to be hanged twice for the same crime), she went on to marry and lived for another fifteen years, no mean feat in mid-seventeenth-century England.

Fears like Tebb's were enough to make some people very wealthy. One such entrepreneur was George Bateson, an inventor who capitalized on widespread taphephobia by pimping out coffins with a modification that became known as Bateson's Belfry. (This modification was only one of a multitude that comprised a small cottage industry of "safety coffins.") In Bateson's model, a small bell was mounted on a coffin's lid, and a string was passed down through the coffin and tied to the hand of the deceased. The rationale, presumably, was that an "apparently dead" person who didn't quite have the strength to lift the coffin lid might still twitch a finger, thereby alerting bystanders that he wasn't truly dead.

The rich and famous could afford to do even more to quiet their anxieties—Duke Ferdinand of Brunswick-Wolfenbüttel, for example, who was terrified of the possibility that he might be buried alive. He crafted very clear instructions so that when he died (really, truly died) in 1792, his coffin was equipped with a window and an air tube. And it was a nail-less coffin, just in case.

Edgar Allan Poe exploited these fears as well in his short story about a premature burial called—wait for it—"The Premature Burial," which was published in Philadelphia's *Dollar Newspaper* in 1844. A man—the unnamed, first-person narrator—is prone to fits of catalepsy that give the impression that he is, in fact, dead. So, naturally, he goes to great lengths to ensure that he is not buried alive.

PRECOXMORTIPHOBIA

It's not difficult to find humor in those anxieties of yesteryear. Certainly some of the more outlandish stories are difficult to take seriously. A coffin with a window? Really?

Nevertheless, the anxieties that were at the root of those stories aren't outlandish at all. If they seem that way to us, perhaps it's only because we've developed a little more trust in medicine and science. We believe—most of us do, anyway—that medicine is good at figuring out which of us is dead. And we can be pretty confident that if there's any chance we're not really dead, someone will notice.

But back then, two hundred years before the EKG had been invented, the hope that you might be resuscitated had to be balanced against the fear that you might not. This phobia—which doesn't actually have a name—is based on a fear that the definition of "death" is changing, and that we're going to be caught on the wrong side. And those concerns are still among us today.

What if, ten years from now, a death due to advanced heart failure will be considered only an apparent death? That's not just idle speculation. Consider this: the cause of death in people with advanced heart failure is often a fatal disturbance in the heart's rhythm. When that rhythm was disturbed fifty years ago, even if the person's heart could be restarted once, his prognosis was grim. Now, though, "sudden cardiac death" is just an indication that a patient needs to have a pacemaker and defibrillator implanted. (We'll meet one beneficiary of this technology in chapter 6.) So what used to be a fatal event is now little more than a nudge toward a new plan of treatment for a patient who can expect to live for a number of years more.

A hundred years ago, the belief that a feather or a horse or a well-placed cigar might restore someone who is "apparently dead" created a sense of hope. However, people paid for that hope with a sense of anxiety that someone they loved—or they themselves—would be mistakenly determined to be dead. In the same way, today's technology is raising similar hopes and fears about whether anything else might be done to prolong a life.

That, in a nutshell, is the modern equivalent of taphephobia. Let's call it "precoxmortiphobia." It's not listed in the DSM-V, but it roughly translates as a fear of a premature death.

There's always another procedure, or another medication, that might be tried. The miracles of resuscitation—when they happen—are so dramatic and sudden that it's difficult indeed not to imagine that they could happen for us, too. And once we begin to consider that, we've opened the door to a profound and inescapable anxiety that whatever death is coming for us is one that can be avoided.

3

The Ice Woman Meets the Strange New Science of Resuscitation

THE ICE WOMAN

When she was twenty-nine years old, Anna Elisabeth Johansson Bågenholm was training to become an orthopedic surgeon in Narvik, Norway. She was smart, talented, and had a bright future. Life was great. And then, suddenly, it wasn't.

On the afternoon of May 20, 1999, Bågenholm and two friends went skiing on a mountain outside of town. They used a route they'd taken many times before, but that particular afternoon, Bågenholm lost control of her skis and tumbled down a steep slope onto a frozen streambed. She landed with enough force to break through the thick winter ice, and seconds later the current dragged her underwater. Only her skis protruded from the opening in the ice.

Those skis saved her from being swept downstream under the ice in what would have been a horrific death. But the force of the moving water against her upper body was so powerful that her friends couldn't extract her. So she hung there upside down, trapped. She was almost entirely underwater and unable to breathe.

Precious seconds ticked by as her friends alternated between frantic cell phone calls and renewed efforts to try to drag Bågenholm from the ice. As those seconds turned into minutes, they thought about giving up, but even after several minutes underwater, Bågenholm continued to struggle. Maybe she'd found a pocket of air under the ice? Something was keeping Bågenholm alive, and she wasn't giving up, so her friends didn't either.

In the meantime, two rescue teams had been mobilized. One was coming from the bottom of the mountain and the other from the top. Soon, though, Bågenholm had stopped struggling and was no longer showing any signs of life.

When the first team arrived, they were unable to free her, and the mood among her rescuers and friends became grim. But they persevered until the second team arrived with a pointed shovel in their toolkit, which they used to hack away enough ice to free her. At that point, Bågenholm had been underwater for eighty minutes. That was considerably longer than the immersion of Michelle Funk, whom we met in chapter 1. What was worse, for at least forty of those minutes Bågenholm had shown no sign of life. So as they pulled her from the stream, her rescuers were not surprised to find that Bågenholm had no pulse, and that she wasn't breathing.

She was also cold. Very cold. And that's where Bågenholm's story diverges from Michelle Funk's. Her body temperature was 13.7 degrees Celsius, which was the coldest temperature that had ever been recorded in a drowning victim. The coldest temperature, that is, for anyone who survived.

Still, she was young and had been healthy. And both her rescuers and her friends—also health care providers—knew that remarkable stories of survival were possible in drowning accidents. So Bågenholm was loaded into the helicopter, where the team placed her on a ventilator and attempted to restart her heart. But they were unsuccessful, and by the time she arrived at the University Hospital of North Norway in Tromsø, about three hours after the accident, the chief anesthesiologist reported that she appeared "absolutely dead." For the record, that's pretty dead.

Still, they didn't give up. (We don't know why they were inspired to persevere, but one can speculate that Michelle Funk's highly publicized recovery might have been in their minds.) Whatever the reason, a rapidly expanding team of physicians and nurses put Bågenholm on a bypass machine that did the work of her heart and lungs. They continued to warm her too. Finally, at 9:15 p.m., almost five hours after her accident, Bågenholm surprised everyone in the ICU when her heart began to beat on its own.

Her subsequent course was rocky, and was punctuated by numerous complications. But soon she began to show signs of improvement, and on May 30, ten days after the accident, she finally woke up. Paralyzed from the neck down at first, and unable to move or breathe on her own, she was nevertheless somehow alive.

Her story was heard around the world. The *Straits Times* of Singapore christened Bågenholm the "Ice Woman"—a name that stuck. CNN, the BBC, and numerous newspapers and television channels trumpeted her remarkable success.

In the end, she made an excellent recovery. Her paralysis resolved and although she had some residual nerve damage that forced her to scrap her plans to become a surgeon, her cognitive function was intact and she was otherwise physically fine. Eventually she went back to work as a physician and went on to marry Torvind Næsheim, one of the friends she'd been skiing with that afternoon.

Bågenholm's story is just one illustration of the remarkable progress that the science of resuscitation has made over the past fifty years. In that sense, her survival, like Michelle Funk's, offers an eyebrow-raising glimpse of what might be possible someday. And hearing her story, you have to wonder what the outer limits of resuscitation might be.

Of course, she was lucky. "I think it's amazing that I'm alive," she admitted on a morning news program. So she was lucky. Very lucky.

Michelle Funk was lucky too. But what sets Bågenholm's story apart is that it has something to teach us. Bågenholm's miraculous survival, and her remarkable physical and mental recovery, can be explained by a theory that related to the one that brought us the Russian Method of resuscitation some two hundred years ago. Bågenholm was a clear beneficiary of that theory, which explains why she survived and how others, maybe, could survive too.

PLUMBING AND WIRING

Before we take a look at the theory that could explain Bågenholm's resurrection, a brief course in anatomy will be helpful. That's why I'm standing in a wide-open space on the second floor of Philadelphia's Franklin Institute. All around me, the cavernous room is scattered with a wide variety of exhibits that beep and pulse and flash. It's like the museum world's answer to Times Square.

One of the prize exhibits here is—improbably—a large fiberglass heart. At about two stories tall, it's scaled to fit inside a 220-foot man. That's large enough to walk through, which many people seem to be doing with the giddy enthusiasm one might expect while about to do something otherwise considered impossible.

Even from across the room, the enormous red-and-purple blob is immediately and unmistakably recognizable as the muscle that keeps us alive. A heart has four chambers that pass blood from the right side of the heart to the left. Blood flows in through one small chamber, the right atrium, which pushes it into a larger, more muscular chamber, the right ventricle. From there the blood goes through vessels in the lungs and then through the left atrium, left ventricle, and out into the aorta. In this fiberglass model, the larger ventricles are clearly visible near the top, as is the massive vein (the vena cava) that wraps around its right side like a pillow. Helpfully color-coded, the structure offers viewers a cheat sheet to figure out what's what. The blue-painted vessels coming back from the body contain blood that is deoxygenated and full of carbon dioxide, and the red-painted vessels denote blood that has been stripped of carbon dioxide in the lungs, and then filled with oxygen. And the centerpiece of this display is the bright red aorta, carrying oxygenated blood out to the brain and all of the other gigantic organs of that 220-foot-tall man.

Alas, today I don't have this magnificent view all to myself. I'm accompanied, it seems, by every second-grader currently enrolled in the Philadelphia school system. The resulting din is overwhelming. On the bright side, the packed masses provide a vivid sense of what it's like to be a red blood cell making your way slowly—painfully slowly—through a heart. Imagine a river of red blood cells each approximately three feet tall, all wearing knapsacks and carrying clipboards, and screaming at the top of their lungs. Which is exactly what these red blood cell reenactors around me are doing.

Tuning them out, at least temporarily, it's possible to get a unique glimpse of what the inside of this fancy machine looks like. And it's at this point I realize with some relief that this refresher course won't take too long. Because however important the heart may be, its mechanics are designed at a second-grade level.

If your heart were a car, it would be that old 1978 MG convertible that your father kept, rusting, in the back of the garage. Like that old machine, the heart has only a few moving parts. And their actions are motivated by a simple logic, and ruled by even simpler laws of physics.

Just as those old MGs were tinkerers' cars, the heart is a tinkerer's organ. Its collection of muscle and nerves and blood vessels is as susceptible to modification and improvement as the engine of a 1978 MG is. And not just susceptible—it's almost as if the heart is *asking* to be patched up, tweaked, and improved. You can almost imagine a cardiac surgeon looking down at an open chest in the same way that a weekend mechanic might peer down into an MG's open hood, thinking pretty much the same thing: "Hey, I bet if I upgraded that wiring, and cleaned out that hose, this baby would run a lot smoother." And indeed, that seems to be the philosophy of two centuries of doctors and scientists who have been trying to patch and fix and improve the heart, as we'll see.

The heart is mostly just a muscle. Inside the giant fiberglass heart, it's a couple of feet thick in places, and its thick cords give the walls a rough texture a little like corduroy. It's solid and comforting. Sort of like being wrapped in a four-ton steak.

As muscles go, the heart is more essential for life than, say, the quadriceps muscle running down the front of your leg that lets you kick a soccer ball. And it's a little more complex, too, as we'll see. But not much.

The heart contracts one time per second, and that's really all it has to do. Like an MG's engine, it just does what it's told to do by a neural center called the sinoatrial (SA) node, which serves as the heart's natural pacemaker. The SA node tells the heart to beat, and it beats. Simple.

The fact that the heart comes from the dealership nicely equipped with this SA node means that it keeps beating without our having to think about it. The beat, as they say, goes on. To appreciate the value of this standard feature, imagine just for a moment what life would be like if we had to concentrate on making our hearts beat in the same way that we need to concentrate on, say, kicking a soccer ball. Now imagine having to do that once every second. Life would be very different. Life would also be very, very short.

The SA node is safely buried between the right and left atria, where blood enters the heart from the body and lungs, respectively. So it's out of sight of the hordes of backpack-toting red blood cells around me. Left to its own devices, and kept out of the reach of meddling second-graders, the SA node produces an electrical impulse about once a second at a comfortable idle. But it can modify its rate, much as a car engine's revolutions can be slowed down or sped up. That impulse travels out to the atria and down through the center of the heart, between the left and right sides. There it reaches the atrioventricular (AV) node, whose not-very-creative name is inspired by its position between the atria and the ventricles. The AV node then regulates each impulse, speeding it up or slowing it down as needed, and sending it on to the heart's ventricles. The impulses go through progressively smaller bundles of nerve fibers known as, in descending order: the bundle of His, the right and left bundle branches, and the Purkinje fibers that carry the signal to muscles.

It takes about 0.19 second for an impulse to travel from the SA node to a muscle fiber. Then the whole process starts over again. The result is a precise choreography that ensures that the atria and ventricles contract in sequence, pushing blood from atrium to ventricle and from the right side of the heart to the left. This choreography is essential. The heart needs carefully calibrated delays so that each part of the heart fires in sequence, in the same way that the cylinders of a car's engine do. For instance, the right atrium needs to contract while the right ventricle is relaxed, because if they contracted at the same time, the larger, stronger ventricle would overpower the atrium, pushing blood backward.

This would cause your cardiac output to drop to zero. Very quickly, you would become unhappy, you would turn blue, and you would pass out. Then you'd be dead. So coordination is a very good thing.

Normally, all this coordination is invisible. But you can see what it looks like if you happen to be looking at an EKG monitor. You can also get a pretty good look if, like me, you happen to be in a two-story fiberglass heart surrounded by pulsing displays.

The heartbeat on an EKG monitor has four main components. First, there's a little bump known as the p wave that indicates the initiation of the heartbeat in the SA node. As this happens, the left and right atria contract, pushing blood into the ventricles. Then there is a long, flat segment—a straight line—during which the signal moves from the SA node through the AV node and down to the ventricles. The length of that segment depends on how much of a delay the AV node is imposing, and whether there are any problems along the signal's path. This is a common place for the signal to get hung up, resulting in varying degrees of a condition called heart block, which is just about as bad as it sounds.

If all goes well, though, next you'll see a series of squiggles as the heart muscle membrane develops a negative charge, and then a positive charge. This is the QRS complex that corresponds to the contraction of both ventricles. Finally, there is the t wave at the very end, when the muscle cells in the ventricle restore their equilibrium and get ready for the next heartbeat. (In the time it's taken you to read this, your heart has gone through fifteen or twenty of these cycles.)

As that electrical impulse moves through the heart, it causes the muscle to contract and relax in sequence. First the atria contract, pushing blood into the ventricles. Then the ventricles contract, pushing blood into the lungs (from the right ventricle) and out into the body (from the left ventricle). Timing is everything. Each chamber has to contract and relax at exactly the right time, so blood flow is coordinated.

If this sounds like a simple system, it is, but only up to a point. The interesting thing about it—and where the analogy to the wiring system in a 1978 MG breaks down—is that the heart doesn't rely only on neural fibers to carry a signal throughout the heart. Unlike the muscles in your leg, for instance, the muscles in your heart form what is called a functional syncytium. In other words, there is a loose aggregation of cells that are all connected to one another, allowing electrical impulses to flow in every direction. This is possible because the muscle cells of the heart don't rely on nerves. Instead, they propagate electrical impulses by way of gap junctions that let signals jump between muscle fibers.

To see why this is important, you have to forget the model of electrical wiring in a car. That wiring goes from point to point, with fuses and switches that can be described in a schematic of lines and boxes. But the heart muscle doesn't really have a schematic. Instead, think of a computer-to-computer Internet network that uses Wi-Fi. Rather than plugging into an access port, a computer in the network grabs a signal from other computers nearby and passes that signal along to others. So as long as someone has a connection, everyone is connected. In the same way, as long as a little chunk of heart muscle is getting a signal, it can propagate that signal to its neighbors. Whereas a signal traveling down a nerve stops if that nerve is severed, if a section of heart muscle is injured, the signal just takes a detour around the affected area.

This system is elegant, and avoids the need for complex wiring to reach every muscle fiber. It also makes the heart more resistant to damage. If some part of the heart muscle is damaged (by a heart attack, for instance), signals can still find their way around. In this respect it's resilient in the same way that a distributed computer-to-computer Internet network in an airport lounge still functions when one person leaves to catch a flight.

There are disadvantages of this arrangement, though. For example, because all heart muscle can conduct electrical activity, a disruption can set up strange patterns in which impulses circle and loop back. This causes heart muscle to contract out of sequence and can lead to sometimes fatal arrhythmias. Indeed, this is the typical cause of arrhythmias after a heart attack. That might happen during the heart attack itself, or in the recovery phase. Still, as a redundant system, in which each backup plan has another backup plan, the heart is really quite well designed.

Next on our tour, there are valves that keep blood flowing in the right direction. Clearly the designers of this 220-foot-tall man's heart had some fun with these, because they're molded with amazing attention to detail. Now my fellow red blood cells and I are passing through the tricuspid valve, which leads from the right atrium to the right ventricle. From the right atrium, we can see three "leaflets," each about the size of a café table. In the body, they are made of connective tissue that flip forward to let blood through, and then snap back to form a seal that prevents blood from moving backward. If these fiberglass leaflets were moving right now, they'd swat us like giant tennis rackets. But they're not moving now, fortunately, and we push our way through, yelling and screaming.

Now we're in the right ventricle, where the most important part of the valve is located. I take a moment to pause and turn around, incurring the wrath of a dozen eager red blood cells behind me. For a moment, I am an unwanted obstacle. I am an adult-size clot.

Soon, though, a red blood cell turns around to see what I'm looking at. From this angle, we can see that each leaflet is tethered to the wall of the heart by perhaps a dozen cords of muscle. Each is as big around as your waist where they leave the wall, tapering to the diameter of your wrist by the time they attach to the valve.

"What are they?" the other red blood cell asks no one in particular.

I explain, over the din, that they're called chordae tendineae. Literally: heart strings.

"Cordy what?"

I tell him that chordae tendineae are strips of connective tissue attached to muscle that keep the valve leaflets from collapsing backward into the right atrium. The mitral valve between the left atrium and left ventricle is similarly equipped. (The valves that guard the exit of the right and left ventricles don't have chordae, and rely instead on rigid cartilage for structure.) Without them, the leaflets would just wave back and forth. Blood would move from the atria to the ventricles, and then back, effectively reducing your blood flow to zero. And then, I add, you'd die.

The red blood cell nods. He gazes around vacantly. Then he lets out a yell, dodges around me, and rejoins the bloodstream.

The beauty of this system is that these valves are mostly passive. With the right timing of contractions, they open in response to increased pressure on one side, and then they slam shut when they're supposed to. That is, they function a little like the exhaust valves of a car's engine, allowing only one-way flow. They're ingenious devices, and researchers have spent decades trying to create facsimiles that work as well (and last as long).

My part of the red-blood-cell parade has now passed through the pulmonic valve and has summited the top of the heart. If we were in a real 220-foot-tall man, we'd be floating out into his 30-foot lungs. There we'd pick up oxygen and release carbon dioxide, moving on through the left atrium, left ventricle, and out into the body. That's one of the trickiest parts of the heart's function, and where accidents often happen. For instance, blood can clot in the smaller vessels, and fluid can leak out into the lungs if the left side of the heart isn't keeping pace with the right.

Perhaps sensing the potential danger ahead, it's at this point that the whooping scrum of little second-grade red blood cells around me becomes densest. And loudest. Then the clot dissolves and we're flowing into the left side of the heart.

As we do, we get an up-close glimpse of the aorta—the largest artery in the body that emerges from the heart and branches repeatedly until blood flows into tiny capillaries. Curious to see how authentic this model is, I look closely at the root of the aorta where it leaves the heart. Sure enough, there are a few small vessels—coronary arteries—branching off and wrapping back around the heart. All muscle needs oxygen to survive, and the heart is no exception. So laced across the outside of the heart is a network of increasingly fine blood vessels that feed oxygen to the heart muscle.

Of all of the pieces of this mechanical puzzle, these little arteries are the ones that are most likely to cause trouble. The valves we just passed through are amazingly durable, as is the neural wiring that surrounds us. But these blood vessels are thin and fragile. They're also not very redundant, meaning they feed a very clearly defined territory. If you develop a clot or a bit of cholesterol plaque in one, it's unlikely that the heart muscle downstream has a plan B. With no other source of blood and oxygen, the heart becomes vulnerable and dies, as do you.

Nevertheless, even these blood vessels are susceptible to tinkering. When they're clogged with blood clots, for instance, they can be cleared with so-called thrombolytic drugs that dissolve blood clots that have blocked off an artery. Balloons inserted on a wire through an artery and then inflated can open a narrowed passage. (Although no respectable cardiologist would explain these interventions this way, you can think of these thrombolytic drugs and balloons as a little like using Drano and a Roto-Rooter, respectively, to open a clogged pipe.) And when all else fails, heart surgery and bypass grafts can route blood around the closed-off sections of blood vessels.

It's so simple, in fact, that it's tempting to try a little tinkering yourself. I mean, how hard could open-heart surgery be? It's just muscle, right?

THE HISTORY OF CHICKEN RESUSCITATION

To say that the mechanics of the heart are simple is true enough. However, that statement ignores the rather important role that the heart plays in keeping us alive.

I'm reminded of this as I watch the man on the operating table in front of me. He's under deep anesthesia, and his chest is wide-open, his rib cage held in place by stainless-steel retractors. But the scariest part of this scene is his heart.

When the heart beats normally in the chest, it's a reassuring metronome, ticking away no matter what physiologic chaos is unfolding around it. But this heart isn't beating. Instead, it's fibrillating—quivering with a frail, fine tremor.

Oddly, no one around the operating table seems the least bit concerned about this heart's lack of activity. They're so nonchalant, in fact, that I'm tempted to point out that this particular heart isn't doing what it's supposed to be doing.

But I don't say anything. I think this is all part of a plan. At least, I hope it is.

The person on the operating table in front of me is a sixty-three-year-old man named Allan. I'm watching his open-heart surgery to get an up-close look at a real heart. About three hours ago, Allan started having crushing chest pain that started in the center of his chest and radiated down his left arm. His wife called 911 and he was rushed to the nearest hospital. An EKG and lab tests found that he was having a massive heart attack. A clot-dissolving drug didn't work, so he was whisked to the cardiac catheterization lab, where dye introduced into his aorta flowed out into the systemic circulation but bypassed his coronary arteries almost entirely. It was as if they weren't there at all.

He had a blockage in his left main coronary artery—the same blockage that killed Joe, the man I introduced you to in chapter 1. From the catheterization lab Allan went straight to the operating room. Now, less than three hours after his chest pain began, his chest is open, he's on cardiac bypass, and he has less than a 50 percent chance of waking up and seeing his wife and grandchildren again.

Allan's cardiothoracic surgeons can't work on his delicate coronary arteries while his heart is still beating. That would be a little like trying to take apart the engine of a 1978 MG while it's running. So with his chest open, and with his blood flowing through a bypass pump, the surgeons stopped his heart by flooding it with a solution that contains potassium chloride. This prevents his heart muscle from conducting a normal rhythm, paralyzing it. That's why no one around me is freaking out about a heart that isn't beating.

Next, in what seems like an eternity but is really just a matter of minutes, they graft a vein to make a detour around the blockage. Then they disconnect the bypass machine and apply a shock to Allan's heart. Allan and his heart get a second chance.

The seconds during which Allan's heart stalled and then roared back to life were probably the simplest part of the entire procedure. It certainly looked simple. Just a quick shock, and all was well.

And yet it took an amazingly long time for would-be resurrectionists to figure out how to pull off this little trick. As far back as the eighteenth century, members of the resurrection club were trying to restart the heart. Some of the earliest research even stumbled on the value of electricity.

Back in 1775, for instance, the Danish physician Peter Abildgaard announced proudly that it is possible to use electricity to resurrect . . . a chicken. "With a shock to the head," he reports, "the animal was rendered lifeless, and arose with a second shock to the chest." That, one would think, would be the point at which most people would stop. But not Abildgaard.

"However," he goes on, "after the experiment was repeated rather often, the hen was completely stunned, walked with some difficulty, and did not eat for a day and night; then later it was very well and even laid an egg." (One can only assume that if Mary Shelley had heard of this poultry reanimation, she would have been unable to resist the temptation to make Frankenstein's monster a Red Bantam.)

Birds seem to have borne a disproportionate share of these early experiments. In another example, from 1796, the German naturalist Alexander von Humboldt was working in his study when a bird collided with a windowpane and fell to the ground. Humboldt ran to fetch a Leyden jar—essentially a homemade battery that stores an electrical charge across a glass-walled container that serves as an insulator. He applied one wire to the bird's beak and—for reasons that are known only to Humboldt himself—inserted the other wire into the bird's anus. The bird, Humboldt reports, was revived for at least a few minutes. Then it died.

Apparently Humboldt was so impressed by this dramatic albeit temporary reversal that he tried the same maneuver on himself, placing one electrode in his mouth and the other in his anus. It's not clear what he hoped to accomplish, given that he had not just hurled himself at a window. However, he reports happily that a shock resulted in "vivid flashes."

Fortunately for all of us, this obsession with shocking chickens and otherwise healthy people was short-lived. Soon science moved on to humans who might benefit. As with much early science, though, a neat chronology is elusive. But the physician and historian Mickey Eisenberg has pieced together a timeline of events that makes as much sense as possible of these early days. Eisenberg places the first possible resuscitation using electricity as far back as 1774, more than twenty years before Humboldt was stimulating his private parts. In that year, the Royal Humane Society's report describes the case of one Ms. Greenhill, who fell out of a window. She was taken to Middlesex Hospital, where the surgeons and an apothecary all shook their heads sadly and declared that nothing could be done to save her. At this point a certain Mr. Squires decided to try electricity. Because, I guess, why not?

Apparently he applied wires to various parts of Ms. Greenhill's anatomy that are probably best left to the imagination. All to no effect. Eventually, though, by design or chance, he applied those wires to her chest, causing her to breathe and eventually to wake up.

The next decade saw other examples, which were summarized by Dr. Charles Kite in 1788, along with a more detailed report of his own adventures with electricity. In that instance, Kite tried the same thing on a young girl who had suffered a fall. The good news is that, as with previous subjects, a shock delivered by a homemade battery seemed to revive her.

The bad news is that Kite's detailed description of his impromptu experiment does not inspire a high degree of confidence that he had even the faintest idea what he was doing. "Electricity was then applied," he reports proudly, "and shocks sent through in every possible direction. . . ." One can only imagine what this scene must have looked like, as Kite pushed his way through a crowd of bystanders and applied wires to any visible skin.

Kite goes on to describe how those shocks caused muscle contractions and how, eventually, the girl woke up. He's often given credit for the first use of defibrillation, but the truth is that it's not at all clear that those shocks actually restarted the girl's heart. It's possible that they stimulated breathing. And of course it's entirely possible that the girl was merely unconscious, and that these shocks accomplished nothing more than a vial of smelling salts would have, albeit with significantly more dramatic flair.

Despite initial advances with chickens and people, these ideas were just a little ahead of their time. Abildgaard and Kite were on the right track, but they hadn't yet figured out how to support the victim's breathing. That science was more than 150 years away, and we'll see those advances in chapter 6. In the meantime, the idea that a heart could be shocked back to life simply sat there, fully charged, waiting until it could be used.

The next big breakthrough came in 1947, when the American surgeon Claude Beck showed that a dose of electricity applied directly to the heart could correct the sort of disorganized, chaotic electrical activity (fibrillation) that is often the last flicker of activity before a heart becomes inanimate. Beck had been working for several years on a defibrillator that could be used during surgery, and he'd even reported two cases where his invention was used. Defibrillation had been successful, but those patients only went on to live for several hours.

But in 1947, Beck's patient was a fourteen-year-old boy undergoing surgery to correct a chest deformity. He was otherwise completely healthy.

That's important because this boy was about to become one of many otherwise healthy patients who led Beck to coin the phrase "a heart too good to die." He meant that this patient was so otherwise healthy that Beck couldn't in good conscience let him succumb to a little thing like a non-beating heart.

As Beck was ending the operation and sewing up the chest, the boy's pulse and blood pressure dropped to zero. As was standard procedure at the time, Beck reopened the chest and performed a manual massage of the heart, squeezing it in a way that should have re-created its normal pumping action.

More than forty minutes passed, with Beck manually pumping away, unwilling to give up. Finally, an experimental defibrillator was brought into the operating room from Beck's lab, and after some trial and error, he succeeded in restarting the boy's heart. Apparently the boy recovered, none the worse for wear.

Allan, by the way, did almost as well. He left the hospital about a week after the open-heart surgery I observed, weak and tired, and unable to walk more than a few steps at a time. But he was alive and grinning the last time I saw him, being pushed in a wheelchair out the hospital's front door, with his wife walking beside him.

To be fair, Allan's procedure was carefully orchestrated. Step by meticulous step, the surgeons and anesthesiologist stopped his heart and restarted it with the proficiency and certainty of trained mechanics. Watching their calm professionalism in the presence of a heart that's not beating, it's tempting to conclude that there's really nothing complicated to it. Just a little shock—that's all you need.

In order to appreciate the magnitude of what the science of resuscitation has achieved, we need to step outside the controlled environment in which Allan's heart was restarted. He had the benefit of an open chest and a heart that had been stopped on purpose as part of a careful, well-choreographed plan. But we need to see what happens when a patient—suddenly and without warning—ceases to be alive and starts being dead.

The "suddenly and without warning" part is going to be a challenge, though. How do we anticipate when and where one of these events will take place? And how can we predict which patient is about to have a very, very bad day? Waiting for a cardiac arrest to occur is a little like waiting for lightning to strike.

Fortunately, there's one place I can think of where I know—with absolute confidence—that someone is going to try to die. I know the time down to the minute when someone's heart is going to stop, and I know exactly when a team of physicians and nurses will descend to try to bring that person back to life. What is less certain is whether they'll succeed.

THE LONG-DISTANCE RUNNER AND HIS VERY, VERY BAD DAY

Allan's cardiac arrest was part of a plan, but for our next patient—Mark—the cessation of his heartbeat was an unwelcome surprise. Mark is lying on the operating table right in front of me, and it is obvious that his day isn't going too well. Which is too bad, because his day started out great.

About three hours ago, this forty-two-year-old scientist and seasoned long-distance runner went into surgery to have a malignant thyroid tumor removed. Following an uneventful procedure, the surgeon closed up the wound, and now the tumor is sitting peacefully on a table not three feet from me. Aside from the fact that he's short one thyroid gland, Mark was otherwise hunky-dory.

Now, however, Mark is not doing so hot. Truth be told, he's dead.

I know this because Mark is not moving, breathing, or performing any one of a number of common and recognizable behaviors that are usually reliable signs of life. I also know this because there's a monitor to my right that displays all of his vital signs, which are conclusively absent. For instance, I'm watching his respirations (nil), heart rate (zero), blood pressure (zip), and EKG tracing (flat). It's a textbook case of someone who is undeniably and incontrovertibly deceased.

And yet, despite the tragically premature death of this forty-two-year-old man, the crew of eight doctors and nurses gathered around the operating table is giggling nervously. An anesthesiologist is trying mightily to force air into Mark's lungs, but those lungs do not seem to be cooperating. A surgeon is unenthusiastically poking at Mark's neck with a scalpel in much the same way that you might poke a beehive with a stick.

Unless something miraculous happens, it's looking increasingly likely that our weekend athlete will never run another marathon. Still, the giggling continues. This, I'm thinking, is going to be difficult to explain to Mark's bereaved family.

Fortunately, that conversation won't be necessary. This isn't a real operating room, and Mark isn't a real patient. I'm actually in the University of Pennsylvania's simulation center, which is designed to re-create the circumstances, confusion, and anxiety of real medical emergencies. All of this—the EKG tracings, the history, and even the trappings of the operating room around us—are the set and props of an elaborate drama that helps OR teams learn to respond to the unexpected.

At the center of this simulation is our patient, known today as "Mark." He's actually a metal and plastic mannequin. His history is fabricated and his physiology is simulated.

However, he's designed with careful attention to detail, so that the team of doctors and nurses surrounding him can do virtually anything they would to a real patient. For instance, when the anesthesiologist inserts a breathing tube into Mark's lungs, a sensor displays the resulting increase in oxygen on a monitor over the operating table. Or a nurse can announce she's drawing blood for a test and, after a suitable interval, the results will be supplied by a disembodied voice. It is a truly amazing arrangement that gives the team real-time feedback, telling them what they're doing well. Or not, as the case may be. At the end of this scene, everyone in the group will critique themselves, and one another, so no one wants Mark to die. But the certainty of his demise, if you ask me, is looking very, very likely.

Suddenly, though, Mark's future begins to brighten. A nurse has wheeled in an automatic defibrillator—a plastic box the size of a milk crate. It's equipped with handy wires that the team attaches to pads placed on Mark's chest. The room is quiet for a moment, and then the defibrillator springs to life.

We all breathe a sigh of relief. It's like everyone's favorite extrovert has just made a grand entrance at a hopelessly dull party. Now, people's expressions suggest, we'll have fun. But it takes only about two seconds for me to wish that this particular guest hadn't been invited.

As soon as it's powered up, the new arrival demonstrates that it has the capacity for speech. (This isn't unusual. Most defibrillators, particularly those used in public settings, provide a computer-generated voice that describes the heart's rhythm, as well as instructions for bystanders.)

Unfortunately for all of us, this defibrillator's voice somehow manages to be both abrasive and dripping with ennui. It reminds me of the disinterested voice that flight attendants use to admonish us—for the millionth time—to use care in opening overhead bins because articles may have shifted during flight. And the defibrillator is using that voice right now to tell the team to "continue CPR."

This advice is greeted with more than a little eye rolling by the doctors and nurses who have been doing exactly that, for the past five minutes with no discernible results.

The defibrillator continues to babble woefully unhelpful instructions. The team continues CPR. And Mark continues to make the gradual transition from temporarily dead to permanently dead.

As the defibrillator drones on, I begin to think of it as Adam, the nom de guerre that Frankenstein's creation bestowed upon himself. With the same exasperating patience, Adam tells the team again and again: "No shock advised." That's because Mark's rhythm is asystole, and there's no impulse that would produce a contraction of the heart. Asystole is essentially a flat line, and this EKG tracing means that the heart is not conducting any meaningful rhythm. And asystole, Adam knows, doesn't respond to shocks.

He knows this, presumably, because he doesn't watch movies. If he did, and if he's an old-enough model, he might have seen the 1990 film *Flatliners*, in which otherwise well-adjusted medical students line up to have their hearts shocked into asystole and then, after varying amounts of drama, back into a normal rhythm. The first part, shocking your way into asystole, at least has some basis in reality. (Although, for the record, this is really not a very good idea.) But once you're there, another shock won't help. You need chest compressions, artificial ventilation, medications, and lots of luck.

This is why we're all listening to Adam suggest again and again in a bored voice that the team should continue CPR. Which they do. Alas, no one has yet found the nerve to tell him to shut up.

As far as I can tell, the only person in the room who is not particularly bothered by Adam's tone-deaf and entirely superfluous advice is Greg Marok, who is standing just to my left. Greg is the simulation coordinator and master of ceremonies for this show, and I'm here as his guest. He's blond, baby-faced, and extra-friendly, doing everything he can to put the simulation participants at ease.

It's Greg's job to orchestrate this simulation in a way that inspires maximum learning—namely by cooking up scenarios that will flummox even the most seasoned teams.

And this team is in danger of being flummoxed, because Mark is still in asystole. Of all of the EKGs that code teams see, asystole is one of the worst because it's very hard to reverse. It's also, not coincidentally, the last thing EKGs show as a patient dies.

What is even more concerning is the fact that Adam seems to have given up. Every thirty seconds or so he's been mournfully telling everyone that there's nothing he can do.

Fortunately for Mark, though, the team is not so pessimistic. Even if Adam can't use the charge he's stored up, the team has a few tricks left. And so, ignoring Adam's gloomy proclamations, they start to do whatever they can.

Since Mark is a mannequin, he won't actually respond to medication. But the members of the team can call out what they'd give a real person, and the computer simulation adjusts Mark's heart rhythm and blood pressure accordingly.

First, they "give" epinephrine. Epinephrine improves the heart's contractility and it also prompts blood vessels to clamp down, which helps to maintain blood pressure. They also give atropine, which blocks acetylcholine receptors, in hopes of lighting a fire under the heart's SA node and getting it to produce a rhythm.

Much to everyone's surprise, the simulation responds and those tricks work. According to the monitor in front of me, Mark's heart is now in ventricular fibrillation. (Fibrillation, which we saw when Allan's chest was open to the world, is the chaotic electrical activity that happens when impulses circle around and around without any recognizable pattern. On an EKG, fibrillation looks like a very fine sawtooth—essentially electrical white noise.) That's still a problem, but it's a better problem to have than asystole, because those little squiggles of electricity mean that Mark's heart is now trying to beat. It's a little like when you're trying to start a car and the engine turns over and then dies. That's bad, but it's a whole lot better than turning the key in the ignition and hearing nothing but ominous silence.

Of course, the onset of ventricular fibrillation has not escaped Adam's keen attention. It seems he's having trouble restraining his enthusiasm. This development injects new life into his electronic voice.

"Ventricular fibrillation!" he announces excitedly. He sounds about as elated as a mechanical voice can sound. It's as if the team has suddenly handed him a task that is worthy of him.

"Stand clear," he warns. If Adam were a real person, he'd be puffing out his chest and strutting forward with self-importance. Now he's the center of attention, and he knows it. He's the man.

The news of ventricular fibrillation seems to breathe new life into the team, too, no pun intended. Now they know what to do, and in fact this is probably something they've done—as practice or for real—dozens if not hundreds of times.

Things start to move quickly. The protocol for ventricular fibrillation begins with three "stacked" shocks, and Adam is delighted to administer the first. Mark doesn't have a pulse, someone announces. And a look at the monitor makes it clear that he's still in ventricular fibrillation—or, in medical-speak, "v-fib." So Adam delivers the second shock, with much the same result, which is to say none at all.

Nevertheless, Adam is still the man. All eyes are on him as he winds up to administer a third shock. He does so with gusto, announcing his intent with a gleeful "Stand clear!"

Then the doctors and nurses descend like birds diving onto a pile of bread crumbs. Mark is still in ventricular fibrillation, and still, there's no pulse. So they follow a carefully choreographed protocol of CPR, shocks, and drugs.

Mark has an endotracheal tube in place, and an anesthesiologist is crouched over him, squeezing a bag that forces 100 percent oxygen into his lungs. Now that the intensive phase of shocks is over, and now that Adam is no longer shouting "Stand clear" warnings, the team can restart CPR. A nurse and a medical student are taking turns doing chest compressions at a rate of about 120 per minute.

The team gives Mark a dose of amiodarone, a drug that stabilizes cell membranes and that is used to prevent or control abnormal rhythms like ventricular fibrillation. It doesn't have any appreciable effect on Mark, so Adam jumps into the fray and administers yet another shock. Then the team follows up with a second dose of amiodarone.

It's at this point that I notice one of the anesthesiologists looking at his watch. At first it strikes me as odd that a physician would be more concerned about an impending lunch date than he is about Mark's future. But then I get it. He's looking at his watch to see how long Mark has been without a pulse. And a brief frown tells me that he's not liking what he's seeing.

It's been—I glance surreptitiously at my own watch—ten minutes. That's ten minutes during which Mark's fictional brain is getting by with no more fictional blood flow than what these intermittent chest compressions provide. CPR is better than nothing, but it's not the same as a normal heartbeat. The anesthesiologist is thinking, probably, that someone had better pull a rabbit out of a hat very soon. And as I look around the room, I notice that others on the team are looking worried too.

Then Adam delivers another shock, which the team follows with a dose of intravenous lidocaine. Lidocaine is one of the most venerable drugs in the crash cart pharmacopeia and is also widely used as a local anesthetic. It works by blocking the channels in a cell membrane that let sodium in.

Since the flow of sodium is a key part of the process that changes the electrical balance of a cell, blocking sodium effectively blocks neural impulses like a pain signal or, in this case, an abnormal heart rhythm.

Sometimes it works. This is not one of those times. But since Adam is eager to administer yet another shock, the team presses on.

Then, during what's become an almost perfunctory post-shock check, one of the nurses notices that Mark's heart has slipped into a sinus tachycardia. That means that a normal (albeit fast) rhythm is back. All around the room, there are expressions of triumph and relief. For a moment, the din is so loud that it's hard to hear Adam's forlorn announcement that "the rhythm is sinus." He sounds almost plaintive, as if he realizes that pretty soon everyone will ignore him and shortly thereafter, he'll be powered down and wheeled back into a closet.

Now Mark has a pulse and a blood pressure. The resistance against the ventilator is signaling to the team that he's waking up. So they give him lorazepam, a benzodiazepine antianxiety drug (like Valium), to calm him down. Then they prepare him for a quick trip to the ICU.

Greg thanks the team and reassures them that Mark has survived. All is well. They're done.

There are high-fives all around as the team heads off to debrief. But poor Adam, lacking arms or hands, is left out of the victory circle. Almost as an afterthought, one of the last nurses out the door reaches over and flicks Adam's power switch as she passes. I swear I can hear a despondent sigh as his lights flicker and then go out.

Everyone's gone except Mark and me, and I'm thinking that he's a very lucky mannequin. He was only "out" for fifteen minutes, so this little interlude without a heartbeat probably didn't hurt him too much. He's still as smart as he was before we started, although, since he's a mannequin, that's not saying much.

For a real person, though, with real family members out in the waiting room, fifteen minutes is an uncomfortably long time. With each passing minute, the cells that depend on blood and oxygen begin to die. The brain is especially sensitive. As more time goes by without a heartbeat, there's a smaller chance that you'd be able to play catch with your son or go horseback riding with your daughter if you come out of it, or that you'd be able to recognize your spouse.

So the science of reviving someone like Mark is only part of the resurrection process. It's one thing to restart a heart. But protecting a brain and other organs until the heart starts beating again is a whole different problem. We've seen how to get a heart working again, but how do we help brains like Mark's to get through incidents like these more or less intact? In other words, how was Anna Bågenholm able to go back to an almost normal life after what she experienced?

The answer, it turns out, is inside our cells. Every single one of them. Figure out what happens inside our cells when they're deprived of blood flow as Mark's were, and you'll figure out how to protect them.

WHAT HAPPENS WHEN CELLS GO BAD

What happens when our vital organs are deprived of blood flow and oxygen? And how can we protect those organs when the heart takes a vacation the way Mark's did? To find out, I'm listening to a man who arguably knows more than anyone else about how our organs respond when the body they're part of is having a very bad day.

Dr. Lance Becker is an internationally recognized researcher in the science of resuscitation, and he looks the part. He's short, balding, and wears square geeky glasses. In fact, if you saw him standing in a hospital hallway, in his plain trousers, open-collar shirt, and thick buttoned shawl sweater, you'd say to yourself: "That's a guy who spends a lot of time thinking about how our innards work." And you'd be right.

Becker studies what happens when cells are deprived of oxygen and when they begin to fall apart. For him, cell innards really only begin to be worthy of our attention when things go very, very wrong. In this respect, he's a little like a forensic psychiatrist who couldn't care less whether you love your spouse but who becomes very interested in you once the police find eighteen decapitated bodies in your basement.

I've heard that Becker is a fantastic teacher, and so in order to hear about his research, I've joined a classroom of medical students who are taking an elective course on resuscitation. Young and enthusiastic, they all want to save lives. And today Becker is teaching them how to do it.

As he tells us about how cells are damaged by a lack of oxygen, it turns out that a cell's descent into pathology is both unbelievably complex and also brutally simple.

The problem, he explains, is energy depletion. Adenosine triphosphate, or ATP, is the main energy source of cells. It's the form in which energy can be stored and used, but the problem is that there's not very much of it in our cells at any given time. So without a constant flow of glucose and oxygen, it vanishes.

The result is cell death, also known as apoptosis. Shortly thereafter, you get person death. This is known simply as death.

But that's just a description of what happens. The question is, how do we slow that process or avoid it altogether? The place to look for the answer is inside the cell. And we're about to get the grand tour.

The first stop on our tour is the cell membrane, which is a thin double layer of lipids (fats) and proteins that keeps a cell's innards inside and the rest of the world outside. One key function of this membrane is to ensure that cells contain the right concentrations of ions, which are charged atoms or molecules. In a healthy, non-dead person, there's more sodium outside cells and more potassium inside. (Each has a positive charge, so their concentrations balance each other.) These gradients are maintained by pumps in our cell membranes, which require energy in the form of ATP. When there's a shortage of ATP, these ions begin to flow freely (more or less). When that happens, they end up in places where they shouldn't be.

Think of the cell membrane as an electric fence surrounding a Kenyan game reserve. The fence keeps people and livestock out, and lions and giraffes inside. And if one or two should slip through, there's a team of park rangers who drive them back to where they belong. But now imagine that the park's budget is tight, so they cut the power to the electric fence. They also fire the park rangers. Now you have giraffes sticking their heads into people's kitchens and hapless cows wandering into the jungle, becoming an ambulatory buffet for the lion population.

As you can imagine, once that starts happening—in cells and in the jungle—everything goes to hell pretty quickly. In the brain, wild swings in ion concentrations prompt neurons to release all of their neurotransmitters in a final kamikaze burst, leading to dramatic increases in unfocused neural activity—essentially many tiny seizures. That's bad.

In the brain and elsewhere, enzymes known as lipid peroxidases break free. These enzymes steal electrons from the lipids in the cell membrane, breaking them down. That's really bad.

Finally, all of these events also lead to problems outside the cell, and particularly in the bloodstream. Out there, inflammation, a lack of oxygen, and general chaos cause platelets to aggregate and clot in small microvessels. This in turn results in further decreases in blood flow, so less oxygen and glucose get into cells. Then ATP becomes further depleted, and more cells die until the organism itself dies. That's about as bad as it gets.

Becker is explaining this descent into anarchy, yet he doesn't seem bothered in the least. In fact, he's grinning. I'm thinking that's because he's got a happy ending hiding in there somewhere. And he does, but it's not exactly what I'm expecting.

As these processes reach a critical point, cell apoptosis pathways are activated. In English, this means that our cells, sensing perhaps that things are looking grim, decide that now would be an excellent time to become dead. And so they check out, leaving someone else to fix things. Or not.

Why is Becker smiling as he's explaining that cells die? It's not exactly a happy ending. But it is. It's what helps us survive a cardiac arrest. And the secret of how that happens can be found in our mitochondria.

Mitochondria are organelles—structures within a cell—that are less than one one-thousandth of a millimeter long. There are a couple of odd things about mitochondria. The first is that they have two membranes, folded over each other like the inner and outer lining of a sleeping bag.

The second is that mitochondria come equipped with their own stash of genetic material. This makes them like cells within cells.

The most interesting thing about mitochondria, though, is that they're responsible for creating ATP. To accomplish this trick, they're equipped with an electron transport chain that uses the energy generated by moving electrons from one molecule to another (usually oxygen) to add a phosphate ion to adenosine diphosphate (ADP). ADP plus this third phosphate produces ATP.

You can think of this conversion of ADP to ATP as refining a fuel into a more potent source of energy. An oil refinery starts with crude oil, which isn't useful for much of anything. But with the addition of energy in the refining process, you get gasoline, which is useful for all sorts of things, like powering a 1978 MG, for instance. So the process adds energy, which is stored in a more concentrated form until it's released.

That analogy also helps to explain why mitochondria close up shop when things aren't going well. In a cell, oxygen drives the conversion of ADP to ATP in the same way that electricity is used to power the processes in a refinery that extract gasoline from crude oil. But if the process is disrupted, then the electron transport chain in mitochondria leaks electrons.

The result is pretty much what you'd see in an oil refinery that's on fire. This is why refineries need a safety switch that can shut down operations to prevent a massive explosion. Trip that switch and the refinery shuts down and volatile products like gasoline stop flowing out, hopefully avoiding a front-page disaster.

Mitochondria seem to have a similar safety switch. They produce ATP when things are good, but when the cell they inhabit loses its supply of oxygen, they shut down. If that causes cell death, that's OK with them because the benefits of that strategy, these little organelles seem to think, outweigh the risks.

Becker is oddly placid about this predilection for self-destruction. More placid than I would be, if all of my mitochondria were thinking about hanging up a GONE FISHING sign. But this impulse to bail out is perfectly normal, he says. It's not pathological. Our mitochondria are just doing what they're programmed to do.

Remember all of the destruction that happens as cells are deprived of oxygen? Chaotic neural activity? Enzymes that dissolve cell membranes? Mitochondria are programmed to avoid all that by shutting down in a controlled way.

Becker isn't happy about this outcome, of course. He knows better than anyone how devastating brain injury can be. No, he's happy simply because these mitochondria are behaving predictably. Their behavior makes sense. And if you're a scientist, that's good news because if you know what a mitochondria is going to do, then at least you have a chance to "reason" with it.

But how?

"What do we do for patients who are in crisis?" he asks the class, changing the subject. "For instance, a patient in cardiac arrest, a patient whose heart isn't beating—what sorts of things do we do?"

"You'd try to restart the heart," offers one student.

"Why?"

"To increase blood flow?"

Becker nods. "What else?"

The group offers a long list of other suggestions. Oxygen? Mechanical ventilation? Drugs to increase blood pressure? Becker nods and smiles.

All of these interventions, Becker explains, give the mitochondria more energy. More oxygen, more blood flow, and a higher blood pressure all deliver more energy to the mitochondria, making them work harder. We're shoveling more energy into cells that can't deal with the energy they've got. We are behaving stupidly.

The problem with resuscitation techniques as we use them now is that they're designed to flood mitochondria with more energy, at precisely the point when those mitochondria are trying their darnedest to shut down. Think back to the refinery analogy. Imagine the moments after a serious refinery accident, when all the safety mechanisms are activated and the plant shuts down. Now imagine wading into the middle of a raging petroleum fire and wrestling the safety valve open.

When Mark the mannequin experienced cardiac arrest, the resuscitation team gave him chest compressions, oxygen, and epinephrine.

In other words, all sorts of things that mitochondria in trouble really don't want. And that, Becker thinks, is a really, really bad idea.

And it gets worse. The damage that physicians can do, despite our best intentions, doesn't stop after resuscitation is successful. In fact, much of what we do in the name of good medical care after we restart someone's heart is just as dangerous and misguided. To illustrate that new danger, Becker hits us with a new analogy.

"Imagine," he says, "that you buy a car. And imagine that, due to carelessness, you let it run out of gas. You trek to the nearest gas station, buy a gallon of gas, and fill it up. Then you turn the ignition and . . . the car explodes." He turns deadly serious, looking out under arched eyebrows for dramatic emphasis. "Would you have bought that car if you knew this was what was going to happen?"

The students around the table are shaking their heads and I follow suit. No. I most certainly would not buy that car.

Becker smiles. He explains that this is a pretty good description of what happens inside a cell after it's been deprived of oxygen. Like a car, our cells run fine when the tank is full. Just as predictably, our mitochondria shut down like a car does when the tank is empty.

But if we try to force fuel into mitochondria that aren't ready to handle it, they're not happy. Like that exploding car, the refueling process is spectacularly disastrous. Not only do we have to be careful about what we put into people during cardiac arrest, we also have to be careful about what we put into cells after we get the heart beating again.

But what's the alternative? We can't just let cells die. We have do something, don't we?

Actually, it turns out that we don't. We don't have to do anything. We can, quite literally, walk away.

THE DEEP-FREEZE MENAGERIE

How can we protect cells—and organs—by walking away? One answer is the one that the Russians thought of two hundred years ago. Cold.

We've known about the benefits of cold for a long time. Longer, even, than we've known what mitochondria are and what role they play in this drama. One of the first—or at least one of the most prolific—researchers to study the Russian Method was Dr. Wilfred Bigelow, a Canadian cardiologist. Initially at least, he wasn't interested in resuscitation, but rather in preserving hearts and brains during surgery.

"It is well known that human beings cannot exist without a functioning heart for much more than three minutes," he says in an early article. This, he continues with typical Canadian understatement, makes heart surgery "a challenge to the surgeon."

No kidding. Imagine stopping the heart, fixing a heart valve or grafting an artery, as Allan's surgeons did, and then putting everything back together again, in three minutes. It's just not possible.

The solution, which apparently came to Bigelow in his sleep one night, was inspired by his experiences in World War II. "I was interested in cooling limbs to see if you could protect them," he explained in an interview. But one night his subconscious took over and he developed far grander plans. "One day I woke up and thought, Why not cool the whole body."

Bigelow was the first to note that dogs could survive fifteen minutes of circulatory arrest if they were chilled to lower than 25 degrees Celsius. In one experiment, thirty-nine "mongrel dogs of medium size" were given an anesthetic and drugs to stop their hearts, then they were chilled to 20 degrees Celsius using a cooling blanket. Next, Bigelow opened the dogs' chests and closed off the vessels leading into the heart with "bull dog clamps," a name whose irony these dogs were probably in no condition to appreciate.

After fifteen minutes, the clamps were released, the dogs were rewarmed, and efforts were made to bring them back. Nineteen of the thirty-nine died. This is a success rate that Bigelow admitted later, in a rare moment of candor, was "disappointing." I'm sure it's particularly disappointing if you happen to be a mongrel dog of medium size.

Then Bigelow had another epiphany. Maybe, he thought, dogs weren't the best test subjects?

He turned his attention to two other species: rhesus monkeys and groundhogs. Monkeys were "more akin to man," and groundhogs, which hibernate during the winter, might withstand even lower temperatures.

Bigelow's experiments with rhesus monkeys followed essentially the same technique that he'd used for the dogs, with a few variations. For instance, after the monkey's chest was opened and the right-side vessels were clamped, the heart was opened as a "token operation." Monkeys went without circulation even longer than the dogs had (up to twenty-four minutes) and at a lower temperature (between 16 and 19 degrees Celsius). At the end, the monkeys were sewn up, rewarmed, and revived.

Of twelve monkeys that underwent a total of thirteen procedures, eleven lived. Bigelow observes, "They were all quite active and well coordinated the day after exposure, and the procedure did not appear to alter their personality or responses in any way."

Of the eleven survivors, most either died of infection as a result of the procedure or were volunteered for further experiments. But one was "saved." Approximately eighteen months after his experiment, Bigelow reported, he was alive and well, and serving as the mascot of a naval station. "He appears normal in all respects," the paper concludes. Except, of course, for the fact that he was a naval monkey mascot. But maybe in Canada that's entirely normal.

Next up were six groundhogs.

Perhaps sensing the lack of potential for narrative drama when discussing groundhogs, Bigelow gets right to the point. First, fully conscious groundhogs were hoisted by their tails and their bellies were injected with sodium pentothal as an anesthetic. Then they were treated just like the monkeys, except that the groundhogs were cooled to between 2.5 and 5 degrees Celsius while their circulation was interrupted for up to two hours.

How did the groundhogs do? Of the six that began the experiment, one died due to hemorrhage, but the remaining five "appeared as well co-ordinated as they were preoperatively." You'll need to decide for yourself just how coordinated an upside-down groundhog might be when it's suspended by its tail and is poked in the belly with a large needle. I'm thinking that the baseline is probably not so good. None of the groundhogs, as near as I can tell, became a mascot of anything.

To be fair, Bigelow acknowledges the risks of hypothermia, an admission that will probably cause the ghosts of an entire menagerie of rhesus monkeys, groundhogs, and mongrel dogs of medium size to nod vigorously in unison. But, ever the cheerful Canadian, he concludes: "We feel, however, that cooling dogs to 20 C with re-warming is a safe procedure." Unless, of course, you happen to be a mongrel dog of medium size, a rhesus monkey, or a groundhog, in which case it's probably safer to stay home.

It turns out that Bigelow was right, and that cold is the secret to survival. Indeed, if anything can rewrite the mitochondria tragedy, it's cold. If cold could protect dogs and groundhogs and monkeys, Bigelow thought, maybe it could protect people, too. Think about Anna Bågenholm's amazing recovery after being pulled from an icy river. But it wasn't until more than fifty years after Bigelow's experiments that the science had evolved sufficiently to prepare the next scene. And that scene played out—where else?—in the original city of zombies.

ZOMBIE DOGS

Early in *Night of the Living Dead*, Johnny and his sister, Barbra, arrive at a spooky graveyard near Pittsburgh to pay their yearly respects to their deceased father. Little did they realize it was the day of the Great Zombie Uprising. Barbra is creeped out by the graveyard and Johnny is making the best of the situation by teasing his sister mercilessly.

"They're coming to get you, Barbra," he says devilishly. There is nervous laughter. Then things get ugly.

Zombies are not real, and Pittsburgh, I'm happy to report, is a very pleasant city. It's full of nice, normal people and entirely devoid of the walking dead. But there was a time not long ago when there really was a kind of zombie there.

Those zombies weren't people, though. They were dogs. Zombie dogs.

At least, that's the unfortunate nickname that a couple of enterprising journalists gave to some very serious and innovative research that was conducted at the Safar Center for Resuscitation Research at the University of Pittsburgh. (The center is named after its founder, Dr. Peter Safar, whom we'll meet in chapter 6.) In these amazing experiments, conducted in the early 2000s, dogs were dead for up to three hours. Then, miraculously, researchers brought them back to life.

Those experiments drew a flood of media attention, which is usually a good thing. In fact, it's at least in part thanks to the media coverage of the Safar Center's previous groundbreaking experiments that the institution is known as the birthplace of CPR.

However, as the researchers at the Safar Center discovered, you really, really don't want to become known as the place that created zombie dogs. And that phrase, certain to be offensive to dog lovers, other scientists, and probably George A. Romero, is the one to which the center's name is inextricably tied. Forever.

Regardless, I'd really like to learn something about these experiments. But I discover almost immediately that if you want to talk to serious researchers about a serious topic like resuscitation, it's not wise to begin a conversation by inquiring about their institution's previous experience with creating zombie dogs. This becomes readily apparent in a telephone conversation with Anita Srikameswaran, the media relations contact for the Safar Center.

"We don't do those experiments anymore," she says with a little more emphasis than is absolutely necessary.

"No dogs," she adds severely, just in case she hadn't been clear.

Then: "None of those experiments."

And that's too bad, in a way, because those experiments offered a rare glimpse of what the future of resuscitation might look like. In the mid-2000s, the Safar Center's team was pursuing a line of research into the applied science of hypothermia that they hoped might eventually help patients. Specifically, they focused on patients with severe trauma and the sort of massive blood loss that might result from a motor vehicle accident, gunshot wound, or injuries suffered on the battlefield, the latter of which were on the rise.

The Safar Center did several experiments along these lines, which were published in respected medical journals. (None of those articles, by the way, said anything about "zombie dogs.") Together, those experiments searched for ways to improve the chances of successful resuscitation through a variety of interventions that include drugs, devices, and temperature regulation.

What did they do? Well, first I should mention that if you're a dog lover, you might want to skip ahead to the next section.

In one experiment, "custom-bred, male hunting dogs" were anesthetized and intubated, and catheters were placed in a vein and in an artery. Then the dogs were taken off the ventilator and blood was removed through a catheter until the dog's blood pressure was 20 mm Hg (normal is more than 100 mm Hg). Finally, the dog's heart was shocked, causing fibrillation. This was done to ensure "zero blood flow," as the article puts it.

Then the researchers packed the dogs in ice and connected them to a cardiopulmonary bypass machine that circulated cold saline solution through the circulatory system instead of blood. In addition, some dogs received oxygen or glucose, or both. The dogs were sustained like this, on bypass but with no ventilation or any other measures, for up to three hours.

This is the point at which things start to get interesting. This is also the point at which any remaining dog lovers should heed that advice about skipping ahead. Seriously.

After spending up to three hours with nothing but chilled saltwater instead of blood, the dogs were rewarmed, and their hearts were restarted. They were also given back the blood that had been stored for them. Thoughtful, I guess.

Then the researchers watched the dogs closely, checking on them every six hours. At the end of the experiment, seventy-two hours later, the dogs that had received oxygen and glucose did the best, and the dogs who didn't receive either did the worst. Of the twenty-four dogs in the experiment, only four were neurologically normal, which is disappointing. But in the glucose/oxygen group, all regained consciousness. Even better, two were normal and the other four had only moderate disability at seventy-two hours. (Although there is no mention in the article of what happened after seventy-two hours, the subsequent description of what the dogs' brains looked like under a microscope should lead savvy readers to recognize that the dogs were eventually euthanized. So we don't know what they would have been like weeks or months later.)

Those experiments led to a storm of media attention, which was both useful and regrettable. Useful because it ignited interest in what was a quantum leap in the science of resuscitation. The ability to revive a dog after three hours without a heartbeat opened the door to a new world of clinical care. If these techniques could be extended to people, accident victims could be stabilized in the field before transport, for instance. And casualties on the battlefield could be airlifted to a well-equipped surgical suite hundreds or even thousands of miles away. All of that was exciting, promising, and awe-inspiring. Or, at least, it should have been.

But a media frenzy was perhaps inevitable, particularly given the rather unfortunate coincidence that the city of Pittsburgh had given birth to cinematic zombies approximately forty years earlier. News reports, for instance, ran with goofy headlines like: "Night of the Living Dogs." The Safar scientists found themselves tossed into the sordid world of tabloid science coverage in which dogs are "hapless pooches" and carefully designed experiments are "unsettling tests."

As if that wasn't bad enough, one article reported that the center's director, Dr. Patrick Kochanek, "angrily denies he's creating a race of zombie dogs fit for a Stephen King novel." That, it seems to me, was sort of a journalistic low blow. It's a little like reporting that Kochanek angrily denies beating his wife. Something any non-wife-beating person would angrily deny.

Predictably, People for the Ethical Treatment of Animals (PETA) weighed in too. Their spokeswoman, Mary Beth Sweetland, declared: "These experiments are indefensible nonsense and the results for humans will be negligible. I would also imagine there are serious consequences for these animals that aren't discussed."

Regardless of what you might think about this particular use of male hunting dogs, Sweetland's dismissal of this line of research as "indefensible" is unfair. And untrue. In fact, these experiments offered pretty convincing evidence that the science of hypothermia could be used to help people. And now it seems pretty likely that Anna Bågenholm survived at least in part because she spent a good portion of her pulseless time submerged in ice-cold water. Whatever your emotional reaction to these experiments, the data were promising and took us a few steps closer to a science that can make miracles like Bågenholm's commonplace.

However, dogs aren't people, and it's a long way from these advances in controlled laboratory settings to eventually helping an accident victim. It's one thing to discover that cold is protective, but quite another to figure out how to cool a gunshot victim on a street corner, or a wounded soldier on the battlefield. To find out how that might work, I have an appointment with someone who thinks an awful lot about cooling people, and pigs.

A PIG NAMED PETUNIA: THE ART AND SCIENCE OF CHILLING

Petunia, I regret to report, is not doing well. I know this because Petunia doesn't have a pulse. Nor is she breathing. In fact, I think it's safe to say that Petunia is no longer alive.

Petunia is what I've nicknamed this example of a *Sus domesticus*. That is, she's a pig. More specifically, she's a pig who is about to teach me something about how it might be possible to cool someone quickly and safely, perhaps saving them in the same way that immersion in an ice-cold stream saved Anna Bågenholm.

To find out how this might work, I'm visiting Petunia and her caretaker, Josh Lampe, a bearded, preternaturally mellow guy who favors soft flannel shirts and worn khakis. He's a bit of an anomaly in the CPR world, because he's not a physician or a biologist, but an engineer. It turns out, though, that if you want to find out how to cool someone, an engineer is exactly who you want on your side.

It's about ten years after the Safar Center experiments and we're in his lab, which is a windowless, low-ceilinged room with bare walls and pristine tile floors. In addition to Josh and me, there are two other people who are wearing gowns and masks and cute hairnets just like I am. One waves a gloved hand. The other nods. This is not a particularly talkative group, it seems, when they're in the middle of an experiment.

I can hardly blame them, though. We're surrounded by racks and racks of computer equipment, whose steady stream of flashing lights and beeps seems to require all of their attention. I also count no fewer than three open laptops that are busily harvesting every piece of data that Petunia is giving us. Imagine the lights and dials and knobs in a 747's cockpit, churned up in a blender and scattered around an average-size hotel room and you'll have a pretty good idea of the sensory overload these guys have to manage.

Petunia is at the center of all of this, lying on her back on a wheeled cart. She's connected to an intricate web of wires and tubes that are monitoring her temperature, oxygen status, and, even at a microscopic level, her mitochondrial function. The crowning element of this setup is an extruded aluminum arch that is placed over Petunia. A motor at the top is driving a three-by-three-inch "fist" that connects with her chest at a rate of about a hundred times per minute. Josh explains that this fist is an automated version of a rescuer performing chest compressions. Unlike a real rescuer, though, this fist doesn't get tired, and it never misses a beat. It will keep going until Josh turns it off. But that won't happen for another hour. In the meantime, Josh and his group are trying to learn as much as they can from Petunia about how to cool a pig and, ultimately, a person.

As Josh is explaining the tubes and wires that monitor Petunia, he emphasizes that he and other animal researchers are monitored almost as closely by various review boards. There are scientific review boards and ethical animal use review committees. This isn't, Josh emphasizes, some Wild West of experimentation, and researchers don't have a cavalier attitude toward the animals they use.

As if in agreement, with each chest compression, Petunia's right forefoot wiggles, just a little. Up-down-up-down. I can't help thinking that Petunia is waving at me.

Now I'm looking back and forth between Josh and Petunia, musing on the wide gap between them. Josh's research is ultimately about saving people, but we're looking at a pig. Why pigs?

When I put this question to Josh, he shrugs in a way that suggests he's heard this question a few thousand times before. "Pigs and people are different," he admits. "But they're actually more similar than you might think." He launches into a technical description of heart muscle and valves that loses me quickly. The point, though, is that despite their differences from humans, pigs are similar enough to make them the best test subjects for studying the way that CPR works.

Josh notes that the biggest challenge isn't anatomy, it's pathology. Or lack thereof. "One big problem is their state of health. Animal models of CPR generally have normal hearts. But most people who have a cardiac arrest are going to have heart disease and clogged arteries. So that's a big difference right there.

"And," he adds, somewhat unnecessarily, "there are anatomical differences in the way they're shaped." He pauses. "Human chests are flat, whereas pigs"—he points at Petunia—"and most dogs, have 'keel chests [chests that protrude].' So it can be difficult to simulate CPR as it's done in humans."

I think for a moment about the dogs I've known. I open my mouth, but Josh is already nodding, as if this is something he's given a lot of thought to. I begin to suggest that there are breeds that have broader, flatter chests, like—

"Dachshunds."

I try to imagine for a moment the reaction of the Dachshund Club of America to the news that their breed has an anatomy that is uniquely favorable for resuscitation experiments. I'm guessing that they would not be overjoyed by that news.

For the purposes of this experiment, though, pigs are just fine. Because today's focus is not on CPR, but on cooling. Specifically, Josh is interested in how to cool a pig—or a person—as quickly as possible down to the temperatures that saved Bigelow's mongrel dogs. But this isn't as easy as it sounds.

It's a matter of physics, Josh explains. You need to find a way to cool a substance—in this case, bacon—as fast as possible, and as evenly as possible. And that, he says, is all about maximizing the rate of energy transfer.

This is difficult because the body is designed to conserve heat and retain energy. Our brains and other vital organs are buried within an insulating shell of water, fat, and muscle (which is mostly water), making our bodies exceptionally good insulators. To an engineer like Josh, you are just a large, soft, ambulatory thermos.

Insulation is great if you're trying to stay warm on a walk to the corner coffee shop on a cold January day. But it's not so great if you've suffered a cardiac arrest and your future brain health—and your ability to pay for a latte with the correct change—depend on getting your brain as cold as possible, as fast as possible. Under those circumstances, you'd wish that your body were a little less insulated.

So efforts to cool victims of cardiac arrest are restricted by our bodies' built-in resistance. This is one of the biggest challenges of improving survival rates. In two randomized controlled trials, for instance, independent groups used cold saline given intravenously to cool cardiac arrest victims. They found a 16 to 25 percent absolute increase in the likelihood of recovering with only minor disability. However, in the real world, outside of a carefully controlled trial, patients may not do as well. In one large study, a patient with an initial rhythm of ventricular tachycardia who gets cold saline would only have about a 50 percent chance of a good long-term outcome.

So if you have a cardiac arrest, you definitely want to get very cold, very quickly. But how? Surely there's something more effective—and quicker—than cold saltwater?

Petunia is just one of a series of pigs that Josh hopes will answer this question. Essentially, Josh and his team are trying to determine how fast they can achieve a target temperature of 32 degrees Celsius. (That's much warmer than Anna Bågenholm's temperature of 13.7, but remember that she was immersed in ice-cold water for more than an hour. And, of course, Josh's work is just a first step.) Actually, the experiments are much more complicated than that, and involve precise measurements of temperatures throughout the brain and body, as well as measurements of blood flow and metabolism. But Josh is wisely giving me the CliffsNotes version.

In these experiments, he's trying several methods, including intravenous cold saline (the current standard of care), a nasal device, and an intravenous cooling solution. In today's session, he's also testing a fourth option. It's a cooling catheter inserted into the femoral vein in the inside of the hind leg and up into the inferior vena cava that runs through the abdomen, where it chills the blood that flows past. Josh and his crew will try these methods on an animal without any circulation, as well as on pigs that are anesthetized, and on those like Petunia who are receiving CPR. All of these variations notwithstanding, the goal of each of these experiments is pretty simple: to see how fast each method cools a pig.

Tracing all of these tubes and wires back and forth from Petunia to the monitoring equipment around her, I'm struck by the vast array of temperature sensors being used. Josh is monitoring temperatures in the brain, in the abdomen, and at the tympanic membrane in the ear, among other places. So which temperature is most important? What cooling rate does he look at most closely?

"Ideally," he says, "we'd cool the brain first, then the heart. So those are some of the most important measurements."

But is that possible? Could you design a cooling mechanism that would *target* the brain?

"Yes, in theory," Josh admits. "You can use a catheter in a carotid artery, which would let you chill the brain quickly. But overcooling is a significant risk." That is, if you bypass the body's insulating properties and feed ice-cold saline to the brain, you could lower its temperature too far, too fast.

I mention that some researchers are trying to do exactly this. As we speak, researchers at the Safar Center are starting a trial that will push the limits of cooling. They'll be trying to chill victims of severe trauma down to 10 degrees Celsius, even colder than Anna Bågenholm was.

Josh nods and gets a dreamy, far-off look. He already knows about that trial, of course. I can't see the portion of his face that's hidden behind the surgical mask, but I think he's smiling. He knows how many lives (and brains) could be saved by advances like that.

Indeed, there are some amazing innovations that have enormous potential. And Josh, an engineer who has devoted his career to trying to save lives, and brains, appreciates them as much as anyone.

Josh's research is part of a race, of sorts. Research is expensive, and the costs of failed experiments are high. But the potential benefits—medical and financial—are enormous. So as with any business that is driven by venture capital and industry, everyone has a dog—or a pig—in that race. And where there is a lot of innovation (and money) there are product names crafted by teams of executives hungry for market share.

The winner in the naming category, at least on my scorecard, is the RhinoChill. This is one of the methods that Josh is testing and whose name at first might seem better suited to the energy drink market than to resuscitation technology. The RhinoChill is a device that is inserted into a victim's nose (hence the prefix "rhino"). It uses liquid perfluorocarbon (or PFC), forced out by high-flow oxygen, to cool the thin skeletal structure known as the cribriform plate that separates your nasal passages from your brain. The liquid PFC is at room temperature, but it evaporates at 0 degrees Celsius. This evaporation pulls heat from the dense network of vessels in the cribriform plate, cooling the blood as it flows through.

Think about the way a dog pants to cool off, using the blood vessels in its tongue to shed excess heat. The RhinoChill is basically a way to get your brain to pant, without the messy slobber on the kitchen floor.

RhinoChill's grammatically unfortunate motto is "Saving Brain. Saving Life," which makes it sound like Tarzan took over their marketing department. Initial reports are promising, though, Josh tells me. Perhaps soon they'll be able to afford a new copywriter.

As the experiment wraps up, Josh tells me more about this and other novel ways of cooling people. Petunia is offering the final bits of data that, one hopes, will guide decisions about which method of cooling works best. As she does, Josh begins to wax philosophical about what he's doing, and why.

"A lot of what we're doing seems crazy," he admits.

It's less a statement than a question. But I shake my head. It doesn't seem crazy to me at all. In fact, it's not difficult to imagine a world in which cooling—rather than CPR—is the first order of business. It seems entirely plausible that cooling could offer more benefit than CPR does.

That was what happened to Anna Bågenholm. At least initially, her body temperature dropped by more than 20 degrees Celsius. But she didn't get CPR until more than an hour later. As Josh's work illustrates, the technical challenges of cooling a person—or even a pig—are considerable. In the end, though, they are just technical challenges, which Josh is convinced we'll solve.

As we're talking, Petunia's mechanical rescuer finally stops. The room is eerily quiet, except for a few frantic beeps from monitoring equipment that have sensed—correctly—that the end is near. Soon those alarms, too, are silenced and the room is still.

4

Science Fiction, Space Travel, and Suspended Animation

THE HUMAN BEAR

If you believe even half of his story, Mitsutaka Uchikoshi is a very, very lucky man. He is also, some would say, not a very bright man. But fortunately those two attributes—luck and lack of common sense—often seem to cancel each other out. At least when we're talking about strange stories of survival.

On October 7, 2006, the thirty-five-year-old Uchikoshi attended a company picnic in a park on Mount Rokko, an oasis in the industrialized Hanshin region of Japan. From downtown Kobe, the park is a ten-minute cable-car ride, which is how Uchikoshi got there with his co-workers. But after a long day of partying, rather than return home the same way, Uchikoshi decided to find his way down the mountain on foot. That probably wasn't his finest decision.

Stumbling down the slope, he became confused and lost his way. He tried to cross a stream, but he slipped and fell, sustaining a pelvic fracture that left him unable to walk. This made his situation particularly dire, because even under the best of circumstances a pelvic fracture can lead to pneumonia, bloodstream infections, and death. And lying on the side of a mountain as night is approaching is hardly the best of circumstances.

So Uchikoshi lay there, exposed to the cold and rain, hoping that some of his friends would miss him and start searching. Soon. But they didn't. Some friends, right?

The night passed, and then the following day. And then another night. Uchikoshi's death was starting to seem inevitable.

Except that it wasn't. Twenty-four days later, a passing climber found him, barely alive. The barely alive part, by itself, is remarkable, given his injuries, exposure, and lack of food or water.

What was truly unusual, though, and what inspired a global media frenzy, was the fact that when he was discovered Uchikoshi's temperature was only 22 degrees Celsius. Temperatures in that range have been recorded for drowning victims pulled from ice-cold water without a pulse, so that's not remarkable. But Uchikoshi was unquestionably alive. His heart was beating, and he was breathing.

That makes his story very different from those of Michelle Funk or Anna Bågenholm. They were dead. They had no heartbeat, and they weren't breathing. Uchikoshi did have a heartbeat, and he was breathing. Yet his temperature wasn't compatible with life.

So he wasn't dead. He was . . . well, no one was quite sure how to describe him. It seemed as though he was in a state of suspended animation.

There's one problem, though. Suspended animation is not possible. Outside of science fiction, people just don't enter a state of suspended animation. Yet Uchikoshi seemed to be doing a pretty good impersonation of someone who had.

Uchikoshi was transferred to the Kobe City Medical Center General Hospital, where he was in critical condition with internal bleeding and a pelvic fracture. But he pulled through, and soon he was talking with the press. Uchikoshi didn't just survive, he also recovered. Doctors were confident about his chances for a full return to normal. For instance, Dr. Shinichi Sato, head of the hospital's emergency unit, said that he believed Uchikoshi's cognitive ability had recovered "one hundred percent."

At the time, there were at least a few researchers who believed that Uchikoshi had managed to achieve a state of suspended animation. They then proceeded to give themselves rotator cuff tears reaching for hyperboles. Uchikoshi's case was "revolutionary," they said. It was "utterly extraordinary."

Was this really an instance of suspended animation? Well, maybe. But we'll never know how many of those twenty-four days Uchikoshi spent in the state in which his rescuers found him. Maybe he'd had a more or less normal temperature for the previous twenty-three days and then, on that final day, he became hypothermic on his way to becoming dead. And maybe that climber found him just before he died.

What is inarguably extraordinary about Uchikoshi's story, though, is the way that it sparked the public's imagination. The media didn't hesitate to point out in headlines that Uchikoshi's survival opened up the possibility of using suspended animation to stabilize patients who were critically ill. "First known human case may help future treatment," was a common theme. These reports didn't just trumpet a miracle, they also promised an entirely new therapy that could buy time for patients who otherwise had no time left.

But could people really benefit from the luck that saved a directionally challenged office worker? Could suspended animation, under controlled conditions, stabilize and protect a patient who is critically ill? And for how long?

SPACE TRAVEL AND LEMON-SCENTED NAPKINS

If you want to think about these questions, and especially if you're a scientist who wants to think about these questions, you'll face one enormous challenge. It's quite difficult to deal with the science of suspended animation without confronting its numerous wacky appearances in science fiction. In dozens of films, suspended animation is used—and dare I say overused—as a plot device.

It's employed to achieve a neat ending as a troublesome character is put into storage (*A.I. Artificial Intelligence*), for instance, and as a convenient setup for the removal of annoying or extraneous characters (*2001: A Space Odyssey*). But its most pervasive use is arguably as an adjuvant to space travel. It seems that you have to be in a state of suspended animation to get anywhere (*Event Horizon*, *Alien*, *Planet of the Apes*, *Avatar*, *Pandorum*, *Outland* . . .). Indeed, this mode of travel is almost inescapable unless you're only going as far as the moon, and sometimes even then (see, for example, *Moon*). Thanks to hundreds of screenwriters over the past fifty years, suspended animation has become the dominant way to bypass the vagaries of economy class space travel.

At least that's a reasonable use for it, and one with which most of us can identify. Think about it the next time you find yourself on a long-haul flight, crammed into an economy class piece of real estate that is barely large enough to provide breathing room for a midsize Chihuahua and her chew toy, next to a screaming child whose parents are eight rows back. How many frequent flier miles would you pay for nine or more hours of suspended animation?

But the real science of suspended animation has taken quite a beating as a result of the humorous uses to which it's been put in film. There is, for instance, Miles Monroe in *Sleeper*, who wakes up after two hundred years, wrapped in tinfoil like last week's corned beef on rye. And there's the French comedy *Hibernatus*, in which the protagonist wakes up after sixty-five years in a small town that has been created in the image of 1905 to avoid the shock that would no doubt ensue were the character to realize what has happened. (The same trope, incidentally, appears in *Captain America.*) And there is my own personal favorite, Douglas Adams's *Hitchhiker's Guide to the Galaxy* series, in which a spaceship's crew puts the entire passenger manifest in suspended animation because they can't bear the thought of taking off without an adequate supply of lemon-scented napkins.

If you're a serious scientist who studies suspended animation, and if you're trying to explain your research to a perfect stranger, the sad truth is that any mention of your work is likely to be met by goofy grins, snide comments, and snarky press coverage. What it is very unlikely to be met with, though, is serious attention, a legitimate scientific reputation, or research funding.

But if suspended animation means slowing a person's metabolism to the point that all of the normal processes of life are put on hold, then similar processes occur in nature all the time. Hibernating animals, like rodents and bears, can achieve a similar state by slowing their breathing (often down to one breath per minute) and heart rate (a couple of beats per minute). A hibernating animal's temperature drops too, often down to just a little above freezing. And the most important feature of hibernation is that the animal's metabolism slows, reducing oxygen consumption to as little as 5 percent of normal.

Reptiles experience a similar process when they "brumate." Since they're cold-blooded, reptiles can't control their own temperature. Therefore, technically speaking, they can't hibernate. But when they brumate, they simply choose an environment in which their body temperature will fall. The net effect of slowed heart rate, breathing, and metabolism is very similar.

Unfortunately, neither hibernation nor brumation are options naturally available to humans. (Humans, that is, except for one Japanese office worker in need of GPS assistance.) But what if it were? After all, animals do it. If we can figure out their secret, then maybe we can teach people to hibernate too.

SPARROWS, "OWZELS," AND RADIOACTIVE GROUNDHOGS: THE EARLY SCIENCE OF HIBERNATION

The question of how animals hibernate appeared as far back as 350 BCE, when Aristotle became curious about how some of our animal friends spend the cold, dark days of winter. For instance, he tells us with ponderous authority that hibernation is widespread among birds. Many, he reports "decline the trouble of migration and simply hide themselves where they are."

Put that way, Aristotle sounds eminently sensible. Indeed, why face the exigencies of air travel? Why not simply take a nap?

Alas, his observations are perhaps overly broad. "And with regard to this phenomenon of periodic torpor," he opines, "there is no distinction observed, whether the talons of a bird be crooked or straight; for instance, the stork, the owzel, the turtle-dove, and the lark, all go into hiding."

I can't comment on the wintering habits of "owzels," whatever they are. But I can report categorically that none of the other birds Aristotle mentions really sleep away the winter. In fact, only one species of bird is known to hibernate: the Common Poorwill (*Phalaenoptilus nuttallii*).

Meanwhile, the swallow, to which Aristotle devotes an inordinate amount of text, here and elsewhere, does not. Aristotle had a strained relationship with birds in general, and swallows in particular. "If you prick out the eyes of swallow chicks while they are yet young," he reports cheerfully, "the birds will get well again and will see by and by." I'd advise taking Aristotle's avian observations with a grain of salt.

Fortunately for science, it seems no one paid much heed to Aristotle's opinions about hibernation. And fortunately for swallows, no one else seemed to think that poking their eyes out was a particularly good idea. Unfortunately, though, no one really thought much at all about hibernation for the next two thousand years. In fact, it's not until the 1940s that people seemed once again to notice that some members of the animal kingdom disappear during winter.

Even then, some of the first "scientific" studies weren't much more helpful than Aristotle's experiments with swallows had been. Consider this gem from an early study of hibernation in hamsters: "That the golden hamster while hibernating is functionally deaf is borne out by the fact that we have never been able to arouse a hibernating hamster by the stimulus of sound." There you have it—the state of much of the science at the time: if you yell at a hamster and the hamster doesn't react, that hamster is deaf.

Soon, though, the science of hibernation itself would wake up. In the 1950s and 1960s, the world began to see a barrage of new research resulting in journal articles with hot titles like "Concentration of urine in the hibernating marmot" and "The unique maturation response of the graafian follicles of hibernating vespertilionid bats and the question of its significance." Its significance is indeed an excellent question, and one that I regret to report no one has yet succeeded in figuring out.

This area of research got a boost in the 1950s, when interest was heightened by a growing realization that hibernation might be protective in the event of nuclear holocaust. In one typical experiment, researchers gave groundhogs a massive dose of radiation and found that hibernating groundhogs lived twice as long as their awake counterparts did. History, alas, is silent about how much this information allayed the pervasive Cold War terror.

Soon an entire generation of researchers would embark on a relentless search for an animal that could help us understand how hibernation works. The list of potential candidates was a long one, though not as long as Aristotle would have us believe. It included bats, woodchucks, marmots, bears, and all sorts of rodents.

Eventually, the scientific community seemed to settle on the humble thirteen-lined ground squirrel (*Spermophilus tridecemlineatus*) as its preferred research subject. This decision was to prove a fateful one for hibernation research. It would also be a turning point for generations of thirteen-lined ground squirrels, for whom life would never again be quite the same.

Thirteen-lined ground squirrels are reliable hibernators. They are also plentiful. No one, it seems, would miss a few who were used for medical experiments. (Note: In the world of research, you really don't want to be a plentiful species.)

Before we get to those experiments, I'd like to meet a thirteen-lined ground squirrel in person. They seem to prefer open grassy areas, like golf courses and cemeteries. But I don't think I've ever seen one. Fortunately, though, I hear about a man in central Ohio who keeps them as pets. So I set off in search of a man and his "thirteen-liners," as they're known in squirrel circles.

CHUCKY THE TRIBBLE

"My wife thinks they're like Tribbles," Joseph admits sheepishly. "You know, those furry little critters on old *Star Trek* reruns?"

He shrugs and chuckles, as if to say, *Women are so silly.* But I see what his wife means. The creature I'm holding in my cupped hands is indeed a bit like the fuzzy little creatures of *Star Trek* fame. But Tribbles were cute, I recall, only until they started multiplying in a way that would position them to take over the universe. That's unlikely with a squirrel, but still, I resolve to maintain a firm grip.

I'm standing in the disused stable of a barn on Joseph's farm. Joseph (most emphatically not "Joe"), is a tanned, clean-cut guy in his sixties who has become fascinated by thirteen-lined ground squirrels and their ability to hibernate. But Joseph isn't a researcher; he's a soybean farmer. Or he used to be. Now he's semiretired and he leases most of his land to a consortium. He's also a squirrel caretaker.

Joseph's fascination with thirteen-lined ground squirrels began one cool September day almost a decade ago when he trapped one for a grandson's science experiment. His grandson lost interest by Christmas, but Joseph was hooked. He kept taking care of the squirrel, and would go out to see it a couple of times every day. He built a bigger cage for it, and then an even bigger one, until finally he had a cage that was larger than many studio apartments.

Joseph explains that he traps all the squirrels he takes care of. "You have to trap them," he explains. "You can't really breed them. Or you can, but it's very, very difficult." They're also asocial and tend to be loners, so he never has more than one in a cage at a time.

If they wander into one of his traps, he says, they must be old and frail. They wouldn't live much longer in the wild. "I think of this as an old-folks home for squirrels," he says. The cage that Joseph proudly shows me is about ten by ten in all dimensions, and it's full of wood shavings, branches, and a pile of hollow logs in a jumble in the corner.

A small metal bowl in one corner is half filled with Purina Cat Chow. According to Joseph, the squirrels love it. He also feeds them insects and berries as a special treat.

OK, so this is really a pretty good life. It's like a squirrel spa.

Joseph has also suspended a couple of large branches from the roof of the cage, presumably as a sort of rodent StairMaster. But none of these is being used right now, because the cage's sole inhabitant, whom Joseph has named Chucky, is in my cupped hands. And Chucky is not in the mood for a workout.

Chucky is curled up tightly in a ball, nose to butt, with his tail wrapped over his head. It's hard to tell, but he looks like he'd be about six inches long, if he were unfolded, not counting his thin, lightly furred tail. In his furled hibernation position, he is approximately the size, shape, and weight of a baseball. Squirrels lose up to one-third of their weight over the winter, and it's December now, so Chucky is probably about halfway down to his dry weight.

Joseph is telling me that Chucky began to hibernate back in mid-October. He'll probably stay asleep, more or less, until April. Males seem to emerge from hibernation first, he says thoughtfully, as if he's not sure what to make of this. Neither am I. But one theory is that since breeding is polygenous, with several males mating with each female, early waking males have the advantage. Alas, Chucky is unlikely to benefit from being an early riser, since he's alone in his cage. He might as well sleep in.

And Chucky seems to be following this plan. Despite the gentle handling that has brought him out of his cage and into my hands, he's not showing any sign of alarm or interest. He is truly out of it.

What's really strange about holding this sleeping squirrel, though, is that he feels . . . cold. It's a little unnerving to hold something that is supposedly alive and to feel no body heat whatsoever. The ambient temperature in the barn is 2 degrees Celsius according to the thermometer on the wall, and I'd believe that Chucky is in that neighborhood. He feels . . . dead.

Yet Chucky is definitely alive. His paws are a bright pink, for instance. And he's breathing, although it takes twenty seconds before I see a breath. He also has a heartbeat, albeit a barely detectable one, which I can feel every five seconds or so. That would be about 12 beats per minute, compared to a normal rate of 200 to 300 beats per minute for a thirteen-lined ground squirrel. I run these numbers by Joseph and he nods.

"Their metabolism winds way, way down. You or I would be dead if our heart rate and breathing got that slow. But these guys . . ." He shakes his head in wonder. "They can stay like this for months."

"Actually," he corrects himself, "they don't stay exactly like this."

And he's right. You'd never guess it from looking at Chucky today, but hibernating ground squirrels do get restless. They don't wake up entirely, but they move around a bit. Joseph discovered this because squirrels like to burrow, and they're happiest if they're covered by a layer of leaves or other debris. Joseph sprinkles their backs with wood shavings from his workshop, so when one is uncovered, he knows that it's been awake.

As my visit wraps up and Joseph walks me back to my car, we talk about how a slowed metabolism might lead to longer life. Sleep for a year and live an extra year? Would that be a good trade-off, he wonders?

Joseph and I are not the only ones pondering this question. In fact, there are a lot of researchers who have been thinking about this trade-off very, very carefully. But they're thinking about a version in which the time asleep is measured in hours and the additional life is measured in years. Specifically, they're wondering whether the trick of hibernation that the thirteen-lined ground squirrel has mastered might someday help people to live for a long time. That story is as strange as any science fiction, and thirteen-lined ground squirrels have a starring role. It starts more than fifty years before Joseph trapped his first squirrel.

SLEEP, SEX, AND SQUIRRELS

Have you ever had one of those restless nights when you just can't seem to fall asleep? When you've tried everything—counting sheep, reading the phone book, doing a crossword puzzle—but nothing seems to work? Well, hope may be on the way. All you need is a glass of warm milk and . . . a tablespoon of squirrel blood. That's the message from a series of experiments way back in the '50s that were designed to find out how squirrels hibernate.

Those studies got off to a rough start. When scientists first began to examine what actually happens to these little guys when they're hibernating, they were nonplussed by what seemed to be a tumultuous physiology. They noted with growing alarm that heart arrhythmias were very common, as were drops in respiration down to a dangerous one to two breaths per minute. These peacefully resting rodents, it seemed, were teetering on the brink of being dead rodents.

Reading this, one begins to worry for the squirrels. Is this normal? Are these little guys OK?

The researchers, too, are obviously concerned about the well-being of their furry friends. To alleviate the anxiety of anyone reading the report, they offer this hearty reassurance: "When an animal in the hibernating state was decapitated, the stump of the corpus showed bright cherry-red blood." So squirrels seemed to be able to survive changes in heart rate and blood pressure that would kill them when they were awake. (They were significantly less well prepared to survive the vicissitudes of being subjects in hibernation research.)

Once researchers figured out that wide swings in heart rate and breathing and blood pressure were normal—and survivable—they wanted to know how hibernation worked. What was it that nudged squirrels into a state that would let them endure these wild physiological changes? And what was it that kept them in this state for an entire winter?

One early theory was that the brain was responsible. Specifically, researchers focused on an area of the brain—the anterior hypothalamus—that they thought might control temperature and metabolism. To test this hypothesis, one researcher by the name of Dr. Evelyn Satinoff placed lucky squirrels in airtight cages. Then she reduced the ambient temperature to 0 degrees Celsius.

"As the oxygen in the container is used up and the concentration of carbon dioxide increases," she reports, "the animal loses its ability to maintain normal body temperature and its temperature falls rapidly." It probably also wonders what the hell is going on.

But once temperature and oxygen are restored, she says, the squirrels recover. Good news, right? Well, it turns out that recovery didn't go so well after Satinoff and her colleagues used electricity to create lesions of the anterior hypothalamus. (If the hypothalamus was responsible for the regulation of metabolism, they reasoned, then damaging that part of the brain would disrupt the body's ability to regulate its temperature.) Indeed, after undergoing this procedure, one poor squirrel (code named CT6) took eleven hours to climb back to his normal temperature (rather than an hour for a normal squirrel), at which point he no doubt heaved a room temperature sigh of relief.

That laborious return to normal earned him a special mention, and even a picture in the resulting journal article. There he is, CT6, in a caption under a photograph. (Unfortunately for CT6, the photograph displays a slice of his brain—a brain that CT6's body was probably not very happy to part with.)

So the brain has something to do with temperature regulation, and maybe it has a role to play in hibernation. Perhaps the brain secretes something that initiates cooling and warming. But what?

The most intriguing answer to that question came a few years later, when a physiologist named Dr. Albert Dawe led a team of researchers to study the hibernation habits of a colony of ground squirrels. During the winter of 1967–68, all the members of that colony of twenty squirrels went into hibernation. The following spring—by March 6, 1968, to be precise—they were awake. All except one.

Already, you may get the feeling that things are not going to end well for this sleeping squirrel. And indeed they do not. If you're a ground squirrel living in a colony of ground squirrels under the supervision of scientists interested in what makes you hibernate, you really don't want to sleep in. You really just don't.

How do I know this? Because in the same paragraph that introduces the lone hypersomnolent rodent, Dawe and his team begin referring to it as "the donor." Even I know enough about scientific experiments to know that if there's one thing you don't want to be, it's a donor.

This particular donor did not fare well. Once all the other squirrels had woken up, bright-eyed and refreshed, Dawe and his colleagues opened the donor's abdominal cavity and extracted three milliliters of blood from its aorta. It was at that point that the poor squirrel decided that it would be an excellent time to check out. So he died.

Undeterred, the researchers took the donated blood and injected it intravenously into two of the donor's buddies, who were placed in a cold room at 23 degrees Celsius along with three other ground squirrels who hadn't been subjected to this surprise transfusion. Then, in a move that would certainly have further unnerved five already confused ground squirrels, the researchers began to watch them very, very carefully. Now at least some of these squirrels might have begun to entertain the notion that they, too, were at risk of becoming donors.

They didn't have to wait long. Within forty-eight hours, the two transfusion recipients fell asleep. And they stayed asleep, more or less constantly, for the next three months. Even more interesting, those squirrels weren't just asleep. Dawe reports proudly that they also demonstrated signs of "true" hibernation, including a balled-up posture, reduced temperature, slowed respiration, and, presumably, growing sexual frustration. (Did I mention that ground squirrels breed in the spring?)

Not content to let sleeping squirrels lie, the researchers began a series of Frankensteinian experiments in which they took blood from the hibernating subjects and transfused it into other squirrels "in a similar way." The report is a little vague about what that means, but one is left to conclude that these transfusions were only made possible by additional untimely deaths for more "donors."

To a discerning reader, this chain of events might bring to mind the sudden deaths of the three unfortunate astronauts in *2001: A Space Odyssey,* which had been released just that year. Early in that film, HAL ends their suspended animation abruptly, irreversibly, and fatally.

Nevertheless, those squirrels gave their lives for what would prove to be the most significant advance in hibernation research thus far. In that daisy chain of transfusions, Dawe discovered something that became known by the acronym HIT: hibernation induction trigger. That is, they'd discovered a massively souped-up version of squirrel Ambien. There was obviously something in the blood of hibernating squirrels that, in the setting of cold, induced hibernation.

Before Dawe could figure out what that substance might be, his experiment came to an end. It was fall, and all of the squirrels started falling asleep naturally, bringing that particular chapter of rodent science to an abrupt end. And none too soon, if you happen to be a thirteen-lined ground squirrel.

So Dawe and his colleagues knew HIT existed. They deduced that it had to exist, given what they'd observed. But they still had no idea what it was. The scientific instruments hadn't yet been invented that would have let them define its chemical composition. That would come later.

PROTECTING HEARTS AND MINDS

Alas, subsequent attempts to replicate some of those early studies have not been terribly successful. Nevertheless, the search for HIT went on. And eventually science began to make some progress.

One of the frontrunners in the scientific search for HIT's identity (a search that, as you may have guessed, has not been kind to the ground squirrel population) is D-alanine-D-leucine-enkephalin. In this nomenclature, alanine and leucine are amino acids, building blocks of proteins. The *D* refers to amino acids' shape, and whether their structure is right- or left-handed (D or L, respectively). Enkephalins are peptides (small proteins) that play a prominent role in the regulation of pain sensation and other functions. They act as a class of opioid receptors (or delta receptors), which are also sensitive to centrally produced endorphins and exogenous compounds like morphine.

I regret to report that this peptide was described, unfortunately and irrevocably, with the melodic but scientifically unprepossessing acronym "DADLE." That medicine has a reliable tendency to turn nouns into active verbs apparently didn't occur to the folks responsible for this particular acronym. While it is routine to "intubate" a patient, to "defibrillate" him, and to "dialyze" him, apparently no one thought ahead to the implications of a medical world in which patients are routinely "DADLEd."

This could be a problem, because a medical future that includes DADLEing is not as improbable as it might seem. Think about it: DADLE seems to protect squirrels from the vagaries of cold, malnutrition, and wide swings in blood pressure and heart rate. Maybe, then, it could work the same magic on humans who are trauma victims or surgical patients?

Some of the earliest hints that DADLE and hibernation could be protective come from experiments that were conducted around the same time that Dawe created his mandatory blood donor program for squirrels. Dr. John Willis, a biologist, was curious about how squirrels' cells continue to function under conditions that really shouldn't support life. In particular, he was interested in their cell membranes.

Willis knew that it takes a lot of energy to maintain the proper gradients of electrolytes inside and outside cells. As we saw in chapter 3 in Dr. Becker's class, the pumps in a cell's membrane need to keep running constantly to make sure that sodium, for instance, stays outside the cell, and potassium, for instance, stays inside. But how well are these energy-intensive pumps maintained during the low-metabolic state of hibernation?

To find out, Willis gathered a group of unsuspecting ground squirrels and hamsters and let them hibernate. Then they were killed, Willis adds in a Professor-Plum-in-the-library-with-a-candlestick rhetorical flourish, "by a blow on the head." Willis then proceeded to chop the squirrels up into tiny pieces.

The hamsters, the report continues, remained awake after the ground squirrels had gone to sleep. This is, of course, very reasonable behavior for any thoughtful hamster who has learned that his squirrel cousins have just been felled by blows to the head, and then minced. But there's no arguing with Mother Nature, and the hamsters, too, eventually went into hibernation. Then they were also converted, immediately and perfunctorily, to rodent mulch.

Once that had been accomplished, Willis found that the gradients across the animals' cells were preserved even in the deepest hibernation. Somehow, even with their metabolism reduced to a tiny fraction of normal, the squirrels' and hamsters' cells continued to maintain themselves at the concentrations to which they'd become accustomed. So the decline in metabolism that hamsters and squirrels undergo seems to spare these essential cellular functions.

That's very interesting, because remember from chapter 3 that one of the chief causes of injury when cells go without oxygen is a failure of their membranes to maintain the proper gradients. Sodium flows in, potassium flows out, and the cell dies. Willis's experiment suggests the possibility that hibernation might stabilize cell membranes, protecting us from the damage that results from a loss of oxygen.

But how might that work? And how might DADLE be involved? To answer that question, we need to take a quick refresher course in cellular physiology.

THELMA AND LOUISE TAKE A ROAD TRIP

As I clawed my way through basic science courses in medical school, I had to memorize dozens of obscure enzyme names. In a desperate effort to keep them all organized in my head, I placed them into two camps. In the social circle of cellular enzymes, it seemed to me, there are some that tend to nudge a cell toward survival and others that are ready to give up when things get difficult.

You can think of these, as I did, as either Thelma or Louise enzymes, from the film of that name. One group (Louise, at least at the beginning of the film) is upbeat, confident, and always up for the next challenge. But the other (Thelma, though not at the end) is moping and doom-saying and generally not much fun to be around.

There's one enzyme in particular, known as p53, which is one of these Thelma enzymes. It has several constructive functions, such as fixing broken DNA. But if its repair efforts are unsuccessful, p53 simply throws its hands in the air and gives up. That is, it nudges a cell toward a graceful exit.

Where this story gets interesting is how in hibernation the balance of power shifts. Conservative (Thelma) enzymes become less influential. Instead, they're replaced by more relaxed (Louise) enzymes.

In a hibernating squirrel, the conservative p53 (Thelma) enzyme drops to approximately one-fourth the concentration found in awake summer squirrels. In contrast, other so-called anti-apoptotic enzymes (Louise enzymes) take over. These are known as anti-apoptotic enzymes because they prevent cell death (apoptosis), and cells produce more of these enzymes during hibernation. Anti-apoptotic enzymes, which go by awkward names like Bcl-X_L and Akt, have the general effect of preserving cell function and preventing cells from dying.

So you've got Thelma enzymes and Louise enzymes on a road

trip together. Normally, the Thelma enzymes are in control. They get to choose when and where to stop, and they keep the radio tuned rigidly to the local NPR station. But gradually those Thelma enzymes become more flexible, open-minded, and cheerful. They loosen up. They start having fun. And they let Louise flip the radio to R&B, for a change.

What's really interesting is that some animals may be able to enjoy the benefits of this shift in enzymes without actually hibernating. All of the fun of a road trip without leaving the driveway. That is, just the ability to hibernate seems to protect cells against hypoxia—the state of being oxygen-deprived.

Some studies have found that even if an animal is awake and alert, as long as it's *capable* of hibernating, it may be better able to survive the rigors of hypoxia during surgery. Most of these experiments were done on—you guessed it—squirrels. Researchers took wide-awake squirrels and subjected them to conditions of hypoxia. Then the researchers euthanized them and removed their livers to examine under a microscope and test those liver cells' ability to function. In these experiments, the squirrels' livers did exceptionally well. At least, they did as well as a squirrel liver can do when it's extracted from the squirrel in which it's been living happily, and placed unceremoniously into a freezer. Even those livers that had until recently been part of a wide-awake squirrel did better than expected when they were frozen. Their mitochondria worked better, for instance. And they produced more bile, and their cells were more viable. (The newly liverless squirrels, though, not so much.)

This is fascinating because it suggests that there may be protective genes in some animals that can be activated by stress. That is, animals may have genes that get switched on very quickly when bad things happen to them, or to their livers. So it's not only the state of hibernation that's protective, but maybe there's a capacity for self-protection that kicks in—or can be made to kick in—when animals aren't hibernating.

This, in turn, leads us to another interesting possibility. Maybe

DADLE can protect organs even when those organs come from animals who don't hibernate. Maybe DADLEing could be therapeutic.

In one fascinating study, researchers removed the hearts of a bunch of rabbits. Usually, the removal of a heart marks the gruesome end of an experiment rather than its beginning. In fact, I think it's safe to say that most organs, once they're no longer connected to the body they've grown up with, will figure out that they no longer have a purpose. They will then cease doing whatever it is they're supposed to be doing.

But a rabbit heart, it turns out, is not particularly perceptive. Detached from its rabbit body, a rabbit heart just keeps going . . . and going . . . and going . . . as the famous tagline says. So it's possible to study rabbit hearts that are out on their own, separated from their respective rabbits.

In this experiment, researchers removed the rabbit hearts and, as a consolation prize, perfused them with a solution that supplied enough oxygen and nutrients to keep them beating. Some of these lonely hearts were flooded with DADLE. Others got the serum (blood minus the red blood cells) from hibernating woodchucks or bears.

What happened? All of the hearts given DADLE functioned better—two times better, in fact—than the other hearts. They contracted with more force, and were able to pump more efficiently. In fact, they were almost normal. That is, except for the rather obvious fact that they were not connected to a rabbit.

So all of this research raised hopes that DADLE—or something similar—might be able to protect the organs of rabbits and rats from injury when they're without oxygen. But what's most exciting about all of this research is the possibility that there might be something out there that could confer some of the protective effects of hibernation in species that don't hibernate—like humans. That possibility might help to explain Mitsutaka Uchikoshi's amazing survival. And more important, it might help to save other lives in the future.

THE RISE AND FALL OF THE "FRENCH COCKTAIL"

Before we think about how we might induce humans to hibernate, we need to think about another, much more basic, question: Why should we bother? Why might human hibernation be a good thing?

Well, as we've seen, hibernation might be a nifty aid to space travel, of course. But there aren't many people who are thinking seriously about that. So I admit that the need for hibernation to reach Mars is not exactly pressing.

However, lots of researchers are thinking very, very seriously about how the tricks of hibernation might be used to help patients. Some researchers think that maybe, someday, hibernation might also be a routine part of clinical care.

The theory—which we've heard before from resuscitation researchers—is that when there isn't much oxygen available, brains and other organs do better if they use less. When the oxygen supply is limited, as it is in the setting of cardiac arrest or major trauma, a brain that needs less oxygen is going to be more likely to wake up. And when it does, it's going to be more likely to retain its ability to control basic functions like walking, talking, and thinking.

As the science of resuscitation has shown us, the quickest way to convince a brain to use less oxygen is to cool it. As a very rough rule of thumb, the brain's metabolic rate decreases by about 6 percent for every 1 degree Celsius drop from a normal 37 degrees. So if you can get a person's core temperature down to 29 degrees, you will have cut his metabolism almost in half. And at 20 degrees, it will be at only 10 percent of normal. So, the logic goes, if we can reduce oxygen requirements this much, it means that people will be able to survive longer under difficult conditions.

Cooling a person who is undergoing CPR isn't so difficult, because that person is dead. Of course there are physics-related challenges, as we've seen. But the person's body is not actively resisting efforts to cool it, for instance, by shivering. And—let's face it—if you go a little overboard it's no big deal because that person is, again, no longer alive.

Hibernation, though, is a whole different ball game. To induce hibernation in people who are not yet dead, you need to cool them while their bodies are fighting back. That is, you need to overcome the body's natural inclination to stay warm. That's an enormous challenge.

The term "artificial hibernation" was first used in 1905, long before we understood much about the mechanisms of hibernation. But it wasn't until the 1950s that science began to make some progress and researchers began to figure out how to cool living people without turning them into dead people.

At the forefront of this movement was Dr. Henri Laborit, a Parisian surgeon with a craggy, leonine face who appears in photographs of the day dangling an elegant cigarette. Laborit was a little obsessed by the Romantic notion that the body can't fight two battles at once. The challenges of maintaining body temperature, he said, compete with the need to fight insults like hypotension and infection. So he came up with the idea of a mix of drugs that would convince the body that temperature regulation wasn't so important after all, allowing it to focus on other things, like staying alive.

This concoction became known as the French Cocktail. Alas, this cocktail is not nearly as much fun as it sounds. Instead of happy-making ingredients like brandy, Champagne, and absinthe, its ingredients were far more prosaic. One was promethazine, which is an antihistamine that is sold today under the brand name Phenergan as an antinausea drug. Another was diethazine, an anticholinergic drug that used to be prescribed for Parkinson's disease, but now isn't used for much of anything.

In fact, it's a little hard to figure out exactly what went into the French Cocktail. The French apparently approached the science of artificial hibernation in much the same way they tackled winemaking. Perhaps subscribing to the terroir theory of drug development, there were multiple versions of the cocktail, each containing as many as six or seven ingredients that varied from hospital to hospital. So it's difficult to tell what those patients were getting.

It's also difficult to figure out whether the French Cocktail improved surgical mortality rates. People weren't keeping reliable records, and reports aren't available. Moreover, even sparse anecdotes and case reports are hard to make sense of, again because patients got many different drugs, in unique combinations, at varying doses.

TAKE ONE GROIN AND APPLY ICE (LOTS OF IT)

Perhaps not surprisingly, given the lack of standardization as well as suspicions of laissez-faire science, the ascendancy of the French Cocktail proved to be short-lived. Soon, pointed questions arose (mostly from across the channel) about its scientific legitimacy. It's at this point that the cutting edge of artificial hibernation research moved to the United Kingdom, where less eloquent minds took over naming responsibilities.

Leading this charge was the Irish anesthesiologist Dr. John Wharry Dundee, who embraced wholeheartedly the value of a reduced metabolism. "So obvious are the advantages of a state of affairs wherein the cellular oxygen requirements are markedly reduced," he says in an influential paper, "that they need not be further discussed." Well, so there then.

Where he and others disagreed with the French, though, was about the value of all the drugs that the French loved so ardently. Dundee argued that the cocktail itself could only be expected to have a modest direct effect on temperature. That is, he didn't think that the cocktail was really cooling people. However, he suggested, if a patient were cooled with ice, drugs might prevent the body from rewarming itself.

Thus he took issue with the initial description of "artificial hibernation," and suggested that this phrase, which had been coming into vogue, was inaccurate. None of the changes induced in the operating room, he said, are physiologically the same, nor are they as profound, as those experienced in the wild. Instead, he began to refer to the process of lowering a patient's temperature and metabolism during surgery as "induced hypothermia with autonomic block." Alas, that hardly sounds like something you'd want to wear your best pearls for, and it doesn't even make for a good acronym.

Faced with this onslaught of science, logic, and boring names, the French Cocktail really never had a chance. It fell out of favor in operating rooms and was replaced by the British "lytic cocktail," which is a step down, nomenclature-wise, if you ask me. ("Lytic," in this case, refers to the ability of drugs to block the autonomic nervous system's natural response to hypothermia.)

Dundee's cocktail was simple, and sported only three key ingredients. Trust the Brits to take all the fun out of experimentation. Those three were promethazine, pethidine, and chlorpromazine. Promethazine, remember, is the antihistamine that made an appearance in the short-lived French Cocktail. Pethidine is better known in the United States as meperidine, an opioid that is sold under the brand name Demerol. It's a mild analgesic but has the unique property of reducing shivering responses. Chlorpromazine is a dopamine antagonist that was one of the earliest drugs used for schizophrenia, and is still marketed under the name Thorazine.

The purpose of that lytic cocktail was not to induce artificial hibernation but to prevent the shivering response that a healthy body exhibits when it realizes that it's getting close to room temperature. Meanwhile, if you really wanted to induce a state of suspended animation—although Dundee didn't use that term—you needed to cool the body. The solution, he proposed, was ice. Lots of it.

In 1953, Dundee and his colleagues reported their success with a series of twenty-six patients who received the lytic cocktail, along with enough ice to chill a couple kegs of beer. First, patients were given the lytic cocktail, and the dose was repeated if the patient showed signs of shivering. What comes next, though, is enough to make me very happy that I didn't need surgery back in 1953.

"Ice bags were first placed on the groin," the report says with admirable British nonchalance. Next, "in the absence of a response" to the ice-on-the-groin maneuver, the body was covered in ice.

Frankly, the absence of a response to a bucketful of ice dumped onto one's exposed groin seems highly unlikely. At least, in anyone who is not already very, very dead. That's British stoicism for you.

But there's more. Dundee's report mentions, as an aside, that no anesthesia was used in these procedures. This, it seems, was a period in which being a patient of Dr. Dundee's was probably only marginally more pleasant than being a donor ground squirrel.

Granted, no one was slicing these patients open and taking blood from their aortas, and all of them truly needed surgery. Nevertheless, they underwent a variety of invasive procedures that today would involve nerve blocks and general anesthesia. For instance, out of twenty-six unlucky bodies, surgeons removed one colon, one breast, two thyroids, two spleens, three bladders, and five stomachs. (Actually, it was three stomachs and two partial stomachs. I rounded up.)

All of this, mind you, without the comforts of anesthesia. And yet, because of the combination of the lytic cocktail and cold (plus perhaps a healthy desire to get the hell out of there), these patients don't appear to have protested. Much.

Using this procedure, Dundee reports proudly that only five patients died, which isn't a bad mortality rate for that time. He also says that temperatures during these procedures sometimes dropped an average of 3.2 degrees, to 32 degrees Celsius.

However, he reports wide variability in those temperatures. The graph of patients' temperature changes over time looks a little like the side view of a pigeonhole mailbox, with a stack of letters all jumbled together but generally leaning to the left. Reading from left to right, some lines start low and remain low. Others start high and drop. Still others float along the top. Statistically speaking, it's a bit of a mess.

Not surprisingly, Dundee's conclusions are correspondingly vague. "One formed the impression," he says, "that the general condition of these cases was better than normal after operation." Those "impressions" might be all right if you're trying to win a beauty pageant, but what one really forms is the hope that more *science* will be forthcoming. Soon. Before one becomes at risk of losing a spleen or half a stomach without the comforts of anesthesia.

A LUCKY MAN

Happily, we've come a long way from placing bags of ice on patients' groins. Many lives and brains have been saved by this progress. And many patients have benefited.

In order to meet one such patient, and his brain, I'm sitting in an office at a mushroom farm outside of Philadelphia. I'm waiting to talk to a man named Thomas, who works here. If anyone can convince me that we've come a long way since the days of the French Cocktail and bags of ice, it's Thomas.

Today Thomas is dressed in rumpled khakis and layers of sweaters to keep warm on a cold March day. He's short and wiry, with the deep tan, leathery skin, and prominent crow's-feet of a man who is used to working outside in all kinds of weather. The one odd item in this picture is a pair of dainty wire-rim glasses that seem out of place on a weathered face that looks like it could have been carved out of a block of gnarled stained oak.

The oddest thing about Thomas, though, is the fact that he's alive. He's alive, and he's sitting in front of me now, because he spent almost an hour in a state of medically induced hibernation. That's an impressive trick, and one that I want to learn more about.

"It all started," he tells me, "when I had a regular checkup with my doctor. And he was listening to my heart and heard a murmur that was new. So I had an echo, and then a CT angiogram. I went in that morning early, had the test, then waited for the results. My doctor came in and asked me how I felt."

He pauses. "I said I was fine. Then he told me: 'No, you're not.'"

The angiogram showed a bulging section at the root of Thomas's aorta, where it emerges from the heart. It was the sort of bulge you see in a balloon if you squeeze it hard enough that a bubble forms. When it comes to balloons and blood vessels, squeezing usually isn't a good idea. What was worse, the angiogram also found a dissection of the descending aorta—a place where it was splitting apart. That's why Thomas's doctor arranged for a helicopter to take him to the biggest cardiovascular surgery center in the region, where he was met by a surgeon who, he hoped, would be able to fix him up.

Thomas turns to a day planner on his cluttered desk, opens it to today's date, and pulls out a sheet of hospital stationery with an elaborate drawing of the upper part of a heart and blood vessels. "I'll never forget; that surgeon drew out for me exactly what he did. He had to place a graft around the root of the aorta, right here, then he needed to patch up this tear farther out here." Thomas doesn't speak as we both look at what was a long procedure.

The problem was time. Even after the chest is open, it takes time to dissect out the heart and blood vessels, and time to excise the damaged blood vessel walls. And still more time to sew a graft into place. For some patients, surgeons can use a bypass machine connected to the aorta to maintain a circulation while they're working on the heart. But in Thomas's case, bypass wouldn't be possible because of the aneurysm in his ascending aorta. So there was going to be a period of as long as an hour when his heart wasn't working and his brain wasn't getting any blood. When general survival and brain health are measured in minutes, an hour is a long, long time.

So what's the solution? How do you get a patient like Thomas through a complex surgery without bypass? And how do you protect his brain so that, when he wakes up, he can go back to his life?

Patients like Thomas are why we should care about the science of hibernation. Humans don't hibernate naturally, because we don't need to survive long, inhospitable winters like marmots do. But we do undergo long, difficult, and complex surgical procedures like the one that Thomas endured. Indeed, if ever there were an inhospitable environment for a brain, an hour without any blood flow is it.

Actually, the hostile environment of the operating room isn't limited to extreme surgery. All but the most minor procedures can be accompanied by wide swings in blood pressure and temperature. For instance, there may be prolonged drops in blood pressure during which the organs that need oxygen most (for example, your brain) aren't getting enough. Add to that the specialized forms of surgery that intentionally cut off blood supply to an organ in order to fix it. Although neurosurgeons are unlikely to explain what they do in such terms, when they repair a brain aneurysm they need to shut off the brain's blood flow in the same way—and for the same reason—that a plumber needs to shut off the water to a bathroom before fixing the toilet.

This need to reduce or eliminate blood flow has put some very strict limits on what surgeons can accomplish. It's also forced surgeons to learn how to operate fast, and to perform increasingly complex procedures within a narrow and inflexible window of time. Surgeons start getting very worried, for instance, when blood flow to a brain is cut off for more than ten minutes. That's not much time at all. In Richard's case, ten minutes wouldn't have been nearly enough time to do the complex repair work that was required.

But Thomas's surgeon told him about another option: a procedure called deep hypothermic circulatory arrest. It's used when a patient like Thomas has a complex surgical problem that will require a long time to correct, and when bypass isn't an option. The heart is stopped, and the brain is cooled, allowing a surgeon to do whatever work needs to be done, hopefully without damaging too many neurons in the process.

This is essentially the same trick that Dundee and his colleagues tried back in the 1950s, but it's much more aggressive. The degree of cooling, for instance, is more extreme. For patients like Thomas, the target temperature is 18 to 20 degrees Celsius. (A normal temperature for a non-dead human, remember, is 37 degrees Celsius.) And then there's the long period—more than thirty minutes—that his brain would be offline.

Overall, this is one of the most complicated and dangerous medical procedures that can be performed. And the risks are correspondingly substantial. The mortality rates are 10 to 15 percent, for instance, and the risk of serious neurologic damage is another 5 to 10 percent. Add to this the risks of any major surgery like blood clots, bleeding, and infection, and you have a procedure that most sane people would avoid if they can.

So it was a big decision Thomas was facing. And he knew that. But then again, it really wasn't much of a decision at all.

"I knew that this was bad. My mom died when she was sixty-six of an aortic aneurysm. Her sister—my aunt—died of the same thing at forty-nine. So I knew it was bad." Bad, I'm thinking, is an understatement. So he decided to go for it.

If the theory is the same as the one that Laborit and Dundee advocated back in the '50s, the details have come a long way. For the entire procedure, for instance, Thomas got the most intensive monitoring possible. The operating team kept a close eye on his temperature, of course. But they also monitored his brain pressure and brain activity using electroencephalography. He also received anticoagulant drugs that prevent blood clotting during cooling and especially during rewarming.

The cooling process, too, has come a long way from the days of French Cocktails and iced groins. For instance, a heart bypass machine cooled Thomas quickly by removing his blood through one catheter, chilling it, and then putting it back into a vein with another catheter. A jacket and helmet with circulating coolant also helped to chill him to the target temperature much more quickly than a bucket of ice would have.

As this was happening, Thomas's blood was removed and replaced—gradually—by saline. Effectively, this diluted his red blood cells to a concentration of less than 50 percent of normal. (If you skip this step and you chill normal blood, it takes on the consistency of a milkshake, with physics that are pretty much the same, except that our arteries and veins are much narrower than a straw.) Those red blood cells were stored so that they could be transfused back into Thomas on the other end of the operation. If he survived.

When his temperature got down to 20 degrees Celsius, the pump was stopped, and the real work started. Thomas's surgeon cut away the damaged aorta and stitched in a new graft of synthetic. Once he was finished and the sutures were tight, the team restarted Thomas's heart, warmed him up, and took him off bypass.

"You know," he says thoughtfully, "it took me a while to bounce back." But he did bounce back. Two months later he was up and around, and his job at the mushroom farm was waiting for him.

As we're talking, I can't help looking for telltale echoes of a brain that once hung up an OUT TO LUNCH sign for an hour. Slurred speech? Memory problems? Unsteadiness? But I'm not seeing anything.

Has he noticed any problems? I ask him. Has anyone noticed anything?

Thomas shakes his head. "No. Well, it took me a while to get back on my feet. But I'm back to work. Everything's OK."

Thomas was lucky. He's lucky to be alive, of course. And doubly lucky to have emerged cognitively unscathed.

But whether he was truly unscathed we'll never know. There's just not much data about subtle cognitive changes that happen when you put a brain through what Thomas's brain was put through. One study asked patients who had undergone a variety of procedures (either with or without hypothermia) whether they'd noticed any changes. Like Thomas, they didn't, and as far as they were concerned, there was no apparent downside of hypothermia.

The problem, though, is you don't know what you don't know. Maybe some of those people *were* a little slower or more forgetful, but they just didn't realize it. Still, it's comforting to know that, at least from the perspective of people who underwent the procedure, they're able to do whatever they could do before.

Thomas feels fine. And he's able to do everything that he could do before all this happened. So he's a success story.

He's not taking it for granted, though.

"I never really thought much about my mortality until this surgery," he says as he's walking me out. "But I think about it every day now. Every day I keep getting older; in a few months I'll be fifty-nine, then I'll be sixty. And then . . ."

Still, Thomas is a living example of what the science of hibernation can accomplish. By cooling him, and by reducing his metabolism to very low levels for more than thirty minutes, Thomas is now alive and well. Even ten years ago, that wouldn't have been possible.

And maybe what was possible for Thomas was only the tip of the iceberg, so to speak. Thomas's brain was protected for thirty minutes. And if thirty minutes is possible, why not an hour? Or two hours?

Maybe eventually it might be possible to put a patient like Thomas into a state of suspended animation for two hours. Or even a day. That's more time than you'd need for even the most complex surgery, of course. But for a soldier injured on a battlefield who's twenty-four hours away or more from an operating room with the necessary equipment, stopping the clock for a day might save a life.

To see how that might work, we can't depend on researchers and surgeons. Nor can we depend on squirrels or groundhogs. Instead, we need help from a whole different team.

TEAM LEMUR

Most of the best early hibernation research has been done on small, furry mammals. Squirrels, especially, have done more than their fair share to advance the science of suspended animation. Indeed, I think it's safe to say that the afterlife is pretty crowded by now with the ghosts of various small critters who deeply regret humans' seemingly insatiable interest in the whole hibernation thing.

However, there is a small but dedicated group of scientists who are convinced that this obsession with rodents amounts to scrambling up the wrong tree. If you want to develop science that will help people, they say, you need to study people. Or, since humans don't hibernate (apart from that one Japanese office worker), you need to find a hibernating species that is as close as possible to humans. I've come to think of this little group of mavericks as Team Lemur.

It's in order to hear their side of the story that I'm standing in a very small, cool room that is lit only by a red light of the sort that you'd find in a photo darkroom. Somewhere in here there are lemurs. Lots of lemurs. And I'm here because lemurs are the only primates—that we know of, at least—that hibernate.

But where are they? I wonder.

Fortunately I won't have to find them on my own. I'm here with someone else whose gaunt form is just barely visible in the reddish glow that surrounds us. And he is someone who knows an awful lot about lemurs.

Dr. Peter Klopfer is one of the founders of the Duke Lemur Center, and he's been semiretired since 2006. But he is still very much involved in the center's activities, and he is a firm believer that studying lemurs—not squirrels—is the best way to glean hibernation lessons that could help people.

Tall, thin, and balding, with an overgrown hedge of a beard that curls back on itself in whorls like an ornamental shrubbery, Klopfer looks like a version of Santa Claus whose wife put him on a crash diet. He speaks intently, yet often seems to be looking off in the distance. Talking with him before we stepped into the lemurs' darkroom was a little like finding myself in a lecture hall full of invisible students. That impression was weakened a little, though, by his decidedly nonprofessorial outfit of running shoes, tracksuit pants, and a T-shirt. In short, I suppose he looks exactly like you'd expect an emeritus professor who studies lemurs to look.

But in our dark little room surrounded by—he promises—lemurs, Klopfer magically assumes an ethologist's mannerisms. He becomes stooped and gawky, and assumes a rapt attentiveness. In the dim light, he looks so much like a primate that I half expect to see a long ringed tail snaking up over his left shoulder. He is the Dian Fossey of the lemur world.

Before we entered the air-locked double doors, Klopfer warned me not to talk once we were inside so as not to disturb the lemurs who are hibernating. So I'm surprised when suddenly Klopfer starts emitting a loud clicking sound that is something like the noise your printer makes just before it spews out a ream of mangled paper and erupts in a puff of smoke.

As Klopfer clicks away there is no discernible activity in the cage in front of us. Then he begins making pursed-lip motions that look like something a female elephant seal might find vaguely romantic. Still, nothing happens.

"This always works," he whispers.

But it isn't working. I mention this.

He squints, looking very intense. Then he grins. Then he points.

A pair of googly eyes appears about six inches from my face. For a second I forget Klopfer's earlier reassurance that the lemurs are in cages. And his admonition not to talk.

In surprise I grunt loudly enough to wake any hibernating lemur. Then I step backward onto Klopfer's foot. He swears.

This visit, I'm thinking, is not getting off to a great start.

But things settle down, and my eyes adjust enough to see three little lemurs hopping around their cage. I'm surprised by how small they are. They're squirrel-size, more or less. However, given the whole squirrel-lemur rivalry thing I'm guessing that Klopfer won't want to hear this.

They're also very agile. One is racing around the top of the cage, upside down, clinging to the wire with tiny humanoid fingers. It moves so quickly and gracefully, it might just as well be right-side up.

And they're very cute. That is, except for their tails, which are big and heavy, swollen with the fat they store there. In fact, the lemur who is currently staring at me through the wire mesh of his cage is equipped with a tail that's almost as big around as he is.

As this little guy and I are watching each other, Klopfer keeps clicking away like a lemur. The lemurs aren't talking back. But they are migrating to our side of the cage and they're paying very close attention to this strange, bearded apparition next to me.

A few minutes later, we've left the lemurs to frolic in the dark by themselves. We're outside in the laboratory and Klopfer is giving me the brief version of the history of lemur hibernation science. He tells me that until very recently, no one thought that primates hibernated. Bears, yes. Rodents, certainly. And, of course, ectotherms like reptiles brumate. But everyone assumed that when primates are faced with a tough winter we just bundle up, turn on the electric blanket, and tough it out.

In 2005, though, a German team of researchers collected the first evidence of prolonged hibernation in fat-tailed dwarf lemurs (*Cheirogaleus medius*). Before then, researchers had only suspected lemurs of hibernating. Although some lemur species had been described as entering a period of torpor, or reduced metabolism, when food supplies were low, no one had caught them in a prolonged period of true hibernation.

The discovery of a hibernating primate opened an entirely new universe of research. It raised the question, for instance, of whether there might be other primates who have been hibernating unnoticed all these years. And it raised the very interesting possibility that other primates—humans, for example—that don't normally hibernate might be able to pull off the same trick.

Before thinking seriously about that question, though, a little family history is useful. It's helpful to know, first, that lemurs are primates, and so are we. But to say we can learn something about hibernation from a lemur is a little like saying I could pick up some running tips from Usain Bolt. Sure, we're both human, but that's pretty much where the similarity ends.

Just as humans and lemurs are both primates, you'd be hard-pressed to arrive at that conclusion just by looking at us. For instance, fat-tailed dwarf lemurs are about the size of a plush squeaky toy you'd toss to your dog. They are technically prosimians, a branch of the primate family that comes from the other side of the tracks. It's obvious that somewhere during the growth of this particular family tree, they just splintered off. So there's a whole lot of DNA distance between a lemur and a person.

And yet, no matter how much they may look like squirrels, they are much closer to us, genetically, than squirrels are. Closer than any other known hibernator is. And that's important, because if we want to understand how hibernation works in a way that might someday help people, it pays to study hibernation in an animal that's as close to us as possible. And that is what Klopfer's lab is all about.

As Klopfer is showing me around his laboratory, he's about to describe the procedures they use. With the squirrel experiment history fresh in my mind, I'm fearing the worst. But since lemurs are some of the most endangered animals in the world, Klopfer's experiments are much more benign.

In all of the experiments that he and his colleagues do, there's nothing that's more invasive than what you or I would endure on a trip to an allergist's clinic. The lemurs are poked and prodded, and very small electrodes are placed under their skin. But no one is cutting these guys open and taking blood from an aorta.

So what *are* they doing? I ask him about DADLE, and Klopfer rolls his eyes and then shakes his head.

"Hibernation," he says slowly, "isn't a conserved trait." He pauses. "It's convergent." What he means, he explains, is that hibernation is something that species have evolved to do in very different ways. Hibernation in a thirteen-line ground squirrel might look like hibernation in a groundhog, a bear, or a Common Poorwill. In fact, to a researcher, the physiology—low temperature, slow metabolism, low blood pressure—looks identical.

However, the mechanisms that make it possible for each animal to hibernate are likely to be very different because these animals have very different genes and physiology. A ground squirrel might arrive at a state of hibernation with a dose of DADLE. Other animals might use other hormones or neuropeptides that we haven't yet discovered. That is, they've all figured out how to hibernate in different ways. Same destination, different paths.

Klopfer makes this case based on the fact that the trait of hibernation is oddly scattered throughout the animal kingdom. There's one bird (the Common Poorwill), and bears and groundhogs and squirrels. And now lemurs. They all hibernate, he says, but they're equipped with very different anatomies and physiologies. It seems highly unlikely that they all hibernate in the same way.

So I ask him about the animals that scientists study in the laboratory. Squirrels are at the top of the list, but there are studies that have tried to induce hibernation-like changes in mice, sheep, and pigs. But Klopfer is, not surprisingly, a strong advocate for Team Lemur.

Klopfer admits that there is a bit of healthy rivalry between researchers who study lemur hibernation and what he describes as "the ground-squirrel team." Each group believes its totem animal offers the best glimpse into how hibernation works.

"It's a friendly rivalry," he clarifies.

I'll bet. The jury's still out, scientifically speaking. The main sales pitch for Team Lemur, though, is humans' proximity to lemurs on the evolutionary tree. If different species get to hibernation in different ways, depending on where they start, it stands to reason that we'll learn most from the species that's starting from a place that's as close as possible to where we are.

Think about it this way: You're standing by a beautiful mountain stream in Montana, holding a fishing rod. On the other side of the stream is someone else, who is also holding a fishing rod, but his fishing rod is connected to a fish. In fact, as long as you've been watching this guy, his fishing rod has been connected to countless fish. So, finally, you ask him which flies he's using and he tells you, reluctantly, that he's using a #10 caddis, a type of artificial fly that's modeled on real insects in the order Trichoptera.

So now you're set, right?

Yes, but only if you happen to have a #10 caddis in your box of flies. If you do, then you're in business. If you don't, then it's a good bet you're going to spend the rest of the afternoon standing by a stream, looking picturesque, perhaps, but holding a fishing rod that will remain annoyingly and persistently unconnected to a trout.

This is the challenge of interspecies research. If we find something that induces hibernation in another species that happens to fit our physiology, then there are enormous opportunities for clinical medicine. But if whatever provokes hibernation in another species doesn't fit with how our bodies work, and isn't even in our physiologic toolbox, then we're back where we started. So what Klopfer and his colleagues are looking for is something in the lemur's repertoire that we also have in ours. But what might that be?

Team Lemur doesn't have a definitive answer yet. But one of the most promising candidates seems to be ghrelin, a 28-amino-acid peptide (a very small protein). Ghrelin is actually a downstream product that results from cleaving a larger, 117-amino-acid peptide (preproghrelin). Another product of that split is obestatin, which Klopfer and his colleagues also suspect is part of the lemur's hibernation mechanism.

The most important thing to know about ghrelin is that it occurs in humans—the equivalent of that #10 caddis fishing fly. As Klopfer tells me this, I can begin to see doors opening to clinical medicine. Figure out how this peptide works in lemurs, he's suggesting, and there's at least a chance that it could work the same way in us.

OK, but how might ghrelin work? The dominant effect—at least as far as we know, Klopfer is careful to clarify—seems to be related to hunger. (The name ghrelin is derived from the Proto-Indo-European linguistic root *ghre*: "to grow.") It has effects on the hypothalamus that induce hunger, for instance. It also reduces satiety, meaning that we eat more before we feel full.

Finally, and perhaps most interestingly, ghrelin seems to enhance reward mechanisms, increasing our responses to pleasurable stimuli, like food (or anything else), including, oddly, alcohol. There's even very limited evidence that intravenous ghrelin can increase food intake in patients with anorexia. Studies like these are a long way from reaching the level of evidence needed for clinical medicine, but they do suggest, at least, that ghrelin is a very real force in human endocrine physiology.

As we walk back across the Lemur Center's campus toward the visitor's center building, I ask Klopfer how ghrelin works in lemurs. Surely it can't just be about hunger regulation and reward behavior. There must be more, right?

Klopfer points out that in lemurs in the wild, the gene that is responsible for making preproghrelin seems to be turned off, or turned way down when days are shorter and temperatures are lower. That reduction, in turn, may reduce metabolism, reduce thyroid function, and may also increase REM sleep. All of that, he says, seems to be a setup for hibernation.

There is one problem, though. Ghrelin levels in humans vary throughout the day. They are also altered by diseases like Prader-Willi syndrome, a hereditary condition. Ghrelin levels may even be reduced after gastric bypass surgery. Despite this variation, though, and despite the fact that some people may have very low ghrelin levels, humans don't hibernate. So if changes in ghrelin in lemurs are associated with hibernation, but similar fluctuations in humans aren't, where does that leave us?

Klopfer isn't sure. This is a new area of research in lemurs, and they just don't have the data yet. They don't even know if ghrelin is the trigger for hibernation. Maybe it's just a side effect. Who knows?

Hormonal pathways are vastly complex, he says. It may be that there are pathways that are activated in lemurs that are dormant in humans, or vice versa. Nevertheless, a better understanding of lemurs, he points out, will at least give us a better sense of what those pathways might be in humans, and where they might be hiding.

HACKING THE HIBERNATION GENE: OF MICE AND MEN

"Don't touch the mouse."

Under normal circumstances, I'm fine with that strategy. Not touching rodents is a pretty good rule to live by. But today I confess that I do want to touch this mouse. Very much.

This particular mouse is lying on an upturned, gloved palm, and it looks about as dead as it's possible for a mouse to look. This is why I want to touch it.

The reason I really want to touch this dead-appearing mouse is because the hand on which it's resting is connected to a researcher named Dr. Cheng Chi Lee. Cheng is a respected scientist at the University of Texas Health Science Center at Houston who is studying suspended animation, or hibernation, or torpor, or, to use the phrase that he's come to rely on as being least likely to incite riots of science-fiction fans: hypometabolism.

Cheng is a short, balding, middle-aged biochemist who has been in the research game ever since he immigrated to the United States in 1986. He has the weathered face, wire-rim glasses, and frumpy clothes of a farmer, or a hardware store clerk, or, I suppose, of a biochemist. He looks eminently trustworthy. And that's a very useful attribute to have if you're trying to convince the world that you just may have stumbled on the secret of suspended animation (sorry, I mean *hypometabolism*).

"This is really neat," Cheng says about the apparently dead mouse.

It's at this point in our brief conversation that I wonder whether, perhaps, Cheng's definition of "neat" is a little different from mine.

When we first walked into Cheng's inner sanctum of hypometabolism research, a hermetically sealed and temperature-controlled room, the mouse that is now lying on Cheng's outstretched hand was scampering around his cage. Then Cheng's graduate student Tre held it carefully by the scruff of its neck and gave it an injection of a very small amount of a clear liquid. Then my little mouse friend was placed in a sealed box in a dark, cool cabinet that allowed us to monitor his activity, temperature, and—most important—his metabolism, from a computer just outside.

The mouse's moniker is #0011, and for the past forty minutes, we've been watching #0011's vital statistics flicker down the screen. The most important of these is his VO_2, a measurement of the milliliters of oxygen per kilogram that he uses per hour. Essentially, VO_2 is a measure of how fast his metabolism is working. Initially, #0011 had a VO_2 of 4,410. That was right after Tre's injection.

For the first minute or so, nothing happened. Then things began to change. Over the next ten minutes, his VO_2 dropped precipitously to around 1,000. Then it dipped into the 900s. Then to 694. Then to 420. Meanwhile, #0011's body temperature went from a healthy 37 to 23.1 degrees Celsius.

Now he's lying on an outstretched palm. And he's dead. Or is he?

I look more closely. As I do, I notice two things. First, he's breathing. Slowly, but he's breathing. So funeral arrangements would be premature.

My second observation takes me a little longer. This little mouse is idling at room temperature, but he's not shivering. And there's no piloerection, meaning the mouse's fur is not standing up to provide increased insulation the way it would if a normal mouse found itself at this temperature.

Is he sleeping? No, that doesn't cover the low body temperature and hypometabolism. Hibernating? No, because mice don't hibernate. None of those terms really fits. He is a mouse that has been injected with . . . something.

I turn to Cheng, who is grinning. He's been waiting for this moment. OK, I ask him. What the hell was in that injection?

A few minutes later, we're in Cheng's office, a Spartan, light-filled space, strewn with papers and a few lonely plants in beakers.

To understand what was in that injection, Cheng says, I need to hear the whole story. It seems like he also wants to give me a sense of how science works and how lucky some of his discoveries have been.

Ten years ago, Cheng was interested in circadian rhythms, the way that our physiology changes throughout a twenty-four-hour cycle, and particularly changes in response to light and dark. In a routine experiment he realized that a co-enzyme called colipase was expressed in tissue where it shouldn't been found. But he found it there only under constant dark conditions. So constant darkness seems to create what he describes to me as a "wonderful process of nature" that hadn't previously been identified.

That unexpected finding prompted Cheng to embark on a hunt for the molecules that might be involved in regulating a gene in response to light. To do that, he used a common technique called high-pressure liquid chromatography. This is a way of dividing a substance into its constituent molecules that can be categorized based on their size and behavior. The result is a little like the "deconstructed" meals you might find on your plate in a restaurant that specializes in molecular gastronomy, in which a common dish is often presented as its component parts.

That analysis pointed to an unexpected nueleotide that Cheng eventually identified as adenosine monophosphate, or AMP. AMP can be broken down to adenosine. Or it can be built up with phosphate molecules to form adenosine diphosphate or adenosine triphosphate, the molecules we met in chapter 3 that are the currency that mitochondria use to create and store energy.

Soon Cheng was able to show that you can induce mouse livers to produce colipase by injecting them with AMP. And it was in one of those experiments that a graduate student told Cheng that mice that had been injected "felt cold." Skeptical at first, but increasingly intrigued, Cheng began to investigate the effects of AMP on physiology in mice, pigs, and dogs.

At this point, Cheng leans back in his chair as if we're done. The answer, he's saying, is AMP. And in fact, the mouse we'd just met—mouse #0011—had received just such an injection before it entered a state of hypometabolism that fooled me into thinking that it was dead.

So AMP causes suspended animation?

Cheng looks at me with the expression of someone who is tempted to say, *If you think that you can summarize ten years of my work with a simple sentence, remind me of why I'm wasting a day talking to you?*

Instead, he switches on his laptop and we huddle in close. He shows me a graph with two lines that travel from left to right above a border that is divided into ten-minute increments. Both lines belong to a mouse after it's injected with AMP and placed in a cold room, as #0011 has been. One line, in black, is the mouse's VO_2. It begins at a plateau and then takes a deep dive almost straight down, so it looks like this mouse's VO_2 dropped by more than 75 percent in fifteen minutes.

The second line, in red, tracks the mouse's temperature. It seems to hover above the black line, following it down, but slowly and hesitantly. The dive in temperature is not nearly as steep, or as deep, as the dive in VO_2 is.

As Cheng takes me through that graph and subsequent slides, there are two things he wants to point out. The first is that we shouldn't attribute too much importance to AMP. It doesn't "cause" hypothermia, he says. At most, it's an initiator of hypometabolism. It sets the stage, but hypothermia is caused by heat loss.

The second point that Cheng wants to make, though, is the most interesting I've heard so far. He points at the graph of the mouse's VO_2, and then at the line of temperature that's lagging behind.

"You noticed?" he asks.

I did. At least, I think I did. They don't coincide. They're not even parallel. VO_2 decreases and then temperature decreases.

Cheng nods enthusiastically. What that means, he explains, is that AMP is causing a drop in VO_2 that's far ahead of—and seemingly out of proportion to—the drop in temperature. VO_2 is decreased, and temperature is decreased, but these decreases seem to be happening almost independently.

That gets my attention, because it reminds me of a study I'd just read. It didn't seem very important at the time, but now I remember it. Cheng's graph has me thinking, improbably, of Yogi Bear.

YOGI BEAR'S TENUOUS MEMBERSHIP IN THE HIBERNATION CLUB

Everyone knows that bears hibernate. Yogi Bear? Jellystone Park? In every other episode, it seems, Yogi is attempting to enter a much-coveted state of hibernation while various malefactors are trying to keep him awake. So you'd think that bears' claim to hibernation is unassailable. But you'd be wrong.

Some of the first doubts about whether bears hibernate were raised by an intrepid naturalist (and bear hunter) named Henry Clapp. He pointed out that when hibernating bears exhale, steam rises from their nostrils. How Clapp got close enough to determine this is a little unclear. However, his vivid descriptions of the insides of freshly killed hibernating bears will reassure concerned readers that he didn't wander into these bear dens unarmed.

Based on his observation of steam and nostrils, Clapp—and many scientists who came after him—concluded that hibernating bears must stay warm. Too warm, they thought, to have the sort of decreased metabolism that is associated with hibernation. Ergo, they don't really hibernate.

It wasn't until more than a hundred years later that it occurred to anyone that even relatively warm bears might still have a reduced metabolism. Asking that question, though, required using a method that was more scientific—and fortunately less dangerous to the scientist—than crawling into bed with them and watching them breathe. But eventually someone did.

In 2010, Oivind Toien and his colleagues rescued a group of five "nuisance bears." (These are wild bears that have proven themselves to be so incorrigible and so prone to menacing human populations that they are euthanized.) Instead, the scientists recruited them into a research study. Spared from execution, the bears happily went to sleep in man-made dens, watched by curious researchers.

What those researchers found at first didn't bode well for bears' prospects in the hibernation club. Although the bears' temperature did drop, it dropped by only 4 or 5 degrees Celsius. Alas, the official scientific report makes no mention of steam issuing from their nostrils.

However, the researchers also found that the bears' metabolism dropped by about 75 percent. That is, their temperatures didn't decrease nearly as much as expected, but their metabolic rate did. Meaning that if you judge by the (old) temperature criteria, bears don't hibernate; if you look at metabolism, they do.

So bears are back in the club, and another important lesson is learned. At least in some species, metabolism can drop way out of proportion to temperature. That may seem like a trivial discovery, but it's not.

Everyone thought that you're not hibernating unless you're *cold* and hibernating. But the discovery that bears can reduce their metabolic rate without a large reduction in temperature means that there might be ways to reduce metabolism that don't involve temperature regulation. At least, not directly.

Remember the French Cocktail and the ensuing debate between Laborit and Dundee? Remember all of those bags of ice placed on all those groins? What if all of that drama isn't necessary at all? What if we could slow metabolism directly? That's an exciting possibility.

There's another reason to be excited that the search for the secret of hibernation is shifting to metabolism. If we're looking for something that alters metabolism specifically, it may be easier to find a trigger that works in both animals and people. Despite all of the differences between mice and bears and lemurs and people, at the level of our cells, we're more similar than different. So if we hit upon something that could reduce metabolism at the cellular level, this could be a game-changer.

Think back to the fly-fishing example, and the guy across the stream who is catching all those fish. I said earlier that if you don't have an exact match for whatever flies he's using, then you're out of luck. But that's not quite true.

What if you don't have to find an *exact* match? What if you happen to have a fly that is universally attractive to all trout, everywhere? Then you don't need a #10 caddis or a #12 coachman or whatever that guy who thinks he's God's gift to trout is using. You can use that all-purpose fly.

Back when I was a resident, whenever I had a day off I'd sneak away and go fly-fishing in some of the small streams in northern Iowa and southern Minnesota. Those fish were a tough audience. They were very picky. And because I didn't know the local streams, most of the time I had no idea what those trout liked.

But I had a trick that would often salvage a fishless day. Most of those streams ran through open meadows, and I knew that those meadows were full of grasshoppers. I also knew that trout like grasshoppers. A lot. Grasshoppers are to trout on a hot summer day what buffalo wings are to a football fan on a Monday night. So even without knowing any of the local patterns or hatches, I'd usually do pretty well with a #12 hopper.

That's what researchers like Cheng are looking for. They aren't hoping for an exact match of a particular animal's physiology with ours. They don't want a cocktail that will mimic all of the complex changes that a hibernating body undergoes. They just want the equivalent of that #12 hopper. They want something that's close enough to trick a body into thinking that it wants to hibernate.

SAM, SMÉAGOL, AND FRODO

As Cheng and I talk about bears and mice, it's starting to sound like AMP might be the biological equivalent of that all-purpose #12 hopper. I'm starting to be convinced. But I still haven't heard a good explanation of how it works. I've seen it induce a state of hypometabolism in a mouse. Or at least that's what it looked like to me. But how?

By way of an answer, Cheng shows me two graphs that are similar to the one we started with. Both have the same shape of the VO_2 curve that drops precipitously and then flattens out. And both show a temperature curve that drops more slowly.

I tell him I still don't get it.

Cheng grins and points at the legends under each. The graph on the right looks almost identical to the one on the left, except that it describes the result of doses of AMP that were ten times greater. The dose of AMP went up tenfold, but the effects were essentially the same.

I think about all of the drugs that doctors prescribe every day, and what Cheng is showing me just doesn't make sense. Take lisinopril, a drug that's used to control blood pressure. Ten milligrams is a typical starting dose. That works for many people. They don't make a 1 mg pill, but if they did, and if you gave it to the average person, it wouldn't have any effect. They don't make a 100 mg pill either, but if they did, and you were dumb enough to give that to another average patient, that patient's blood pressure would probably drop precipitously, causing dizziness, lightheadedness, and maybe a loss of consciousness. That's a typical dose-response relationship. More drug leads to a greater effect, up to the point of toxicity and even death.

Cheng understands the oddness of his results too. "The first time I saw this, I thought there was a mistake. It couldn't behave like this. It's just wrong." He sounds like he was seriously offended that his AMP isn't behaving in the way that it should.

It wasn't wrong, though. In fact, the lack of a dose-response curve offers a couple of clues as to how AMP works. It's what is called an allosteric reaction.

An allosteric reaction, Cheng explains, has three "actors." There's a protein. Then there's whatever molecule the protein usually binds to. Then there's a third actor—call it X—that binds to the protein. That third actor, X, changes the affinity of the protein for the molecule to which it normally binds. Depending on whether X increases or decreases that affinity, it's described as either a positive or negative allosteric reaction.

Think about two people who don't really get along. Call them . . . Sam and Sméagol. Left to their own devices, they'd never spend any time together. But then add a third person. The X factor. Let's call him Frodo. With the addition of Frodo to the mix, Sam and Sméagol manage to tolerate each other, at least for the duration of their journey to Mount Doom. They're never friendly, and they never really get along. But they coexist. That's the way a positive allosteric reaction works.

Allosteric reactions tend to produce results at predetermined levels. A protein by itself will have a certain attraction for another molecule. If you add X, the protein's attraction for that molecule is kicked up a notch, or down a notch, but only within a predetermined range. Just as there's no way Frodo will ever convince Sam and Sméagol to become BFFs, an allosteric reaction can't convince a protein to bind more tightly to a molecule than it's programmed to.

So—and here's the most important point about allosteric reactions—you get the same change in attraction and binding whether you add X at a 1 percent concentration or a 10 percent concentration. Sound familiar? That was exactly the result that Cheng was seeing.

Even better, once Cheng figured out AMP was part of an allosteric trio, he knew there aren't that many allosteric reactions in the body. So he could make an educated guess about how AMP works. That's because he knew about an interesting and very well-known negative allosteric reaction that happens in red blood cells. A molecule called 2,3-diphosphoglycerate (known either as 2,3-DPG or BPG, for biphosphoglycerate) reduces the attraction of hemoglobin for oxygen. In the presence of BPG, hemoglobin is slower to bind to oxygen, and quicker to release it to tissues that need it. So at a given pressure of oxygen in the blood, adding BPG means that less of that oxygen will get bound to hemoglobin.

The key point here is that the interaction between BPG and hemoglobin is a negative allosteric reaction. If you expose red blood cells to a little BPG, you reduce their capacity to carry oxygen. Expose them to a lot, and you get the same effect.

The most important implication of this result is that it pinpoints the location of AMP's effects. "Now," Cheng announces, "we think we know where this happens."

We do?

"Where is BPG?" He looks at me steadily.

I'm stumped.

"Where is it . . . in the body?" he clarifies, helpfully. Thanks for that.

Well, red blood cells, obviously, I tell him.

"And?"

I stretch my memory back to the first year of medical school, without much success, but then it hits me.

Nowhere else?

"Right!" Cheng says. (Actually, it also appears in the placenta, he admits. But I'm pretty sure my male mouse didn't have one of those.)

This means that ordinary red blood cells might have a role in hibernation. As we've seen, people have looked to the brain and the liver and even the eyes, if you count Aristotle's strange fascination with swallows. However, no one was really thinking about red blood cells.

But how does it work?

The answer to that question isn't complicated, but getting to it requires a basic understanding of how our bodies break down glucose (sugar) to produce energy. This process, known as glycolysis, is the way a cell uses the energy that is stored in glucose to create adenosine triphosphate, or ATP—the energy molecule we met in chapter 3. Glycolysis has two interesting features that are relevant to Cheng's work.

First, remember that adenosine comes in three forms. There's adenosine monophosphate (with one phosphate molecule). And there's adenosine diphosphate (with two) and adenosine triphosphate (with three). When you add AMP to red blood cells, it becomes ADP by pulling a phosphate molecule off ATP.

That, in turn, leads to a cellular shortage of ATP, which slows metabolism. Our cells need an initial investment of energy to make energy, and in the same way that a tough economy restricts the availability of startup capital and makes it more difficult to get a loan, a shortage of ATP makes it that much more difficult for cells to produce new ATP. Indeed, Cheng and his colleagues have found that

AMP-treated blood cells have more glucose than they should, which suggests that they're not breaking it down at the usual rate. That is, AMP seems to slow down metabolism.

It's the second feature of this process, though, that gets us back to BPG. When there's a shortage of ATP, then BPG—normally a by-product—is suddenly produced in large amounts. And that happens—if Cheng is right—in every red blood cell in the entire body. That means that a very focused change would induce a whole-body hypometabolic state. By simply shifting the oxygen-hemoglobin binding curve a bit, AMP—if indeed it's AMP that's responsible—could force a systemic downregulation of metabolism. That's one change, to one molecule, that then affects every organ in the body. Amazing.

It's a little like using gas prices to change people's behavior. You can do all sorts of things to encourage people to walk more and drive less. But as a single, global intervention, there isn't anything that works quite as well as tripling gas prices at the pump. So maybe that's AMP's role—to create a tax on oxygen that means that cells everywhere need to conserve.

But is that safe? It seems a little risky. Surely there's a potential downside to decreasing oxygen supply. Like brain damage. Or death. But Cheng smiles.

"It's perfectly safe, as long as you keep ambient temperature low so that the body temperature can drop. It's the cooling of the body that allows for the decreased need for oxygen. With a low enough temperature . . . no problem."

THE BADASS BEAGLE

In order to reassure me that what he's doing is safe, Cheng fires up a video to prove his point. He points at the screen.

"This is a dog," he announces.

Indeed it is. He has floppy ears and wiry fur. He looks like a beagle having a bad hair day, and he's very cute. I find myself thinking back to experiments with thirteen-lined ground squirrels and "mongrel dogs of medium size," and I'm fearing the worst.

First the dog is given a haircut, to reduce his ability to keep warm. That's not too bad. There are a few tufts of fur around his ears that are sticking up at 45-degree angles, but this just makes him look a little badass. At least, as badass as a beagle can look. Then he's injected with AMP.

The video was taken in a lab in China by one of Cheng's collaborators, and he admits they weren't sure of the optimal dose of AMP for a dog. So they made a guess based on their work in mice.

He shrugs. I wince. This, I'm thinking, is not going to end well.

But the dog seems unperturbed about his immediate future health prospects. He sits there panting and flopping his ears back and forth on Cheng's laptop screen. As we're watching, the dog stumbles around a little. Like an undergraduate at a frat party. Then the dog passes out, as gradually and naturally as if he were falling asleep.

I'm worried. But it seems that Cheng isn't. And he reminds me why there's really nothing to worry about.

Because the relationship between BPG and hemoglobin in the presence of AMP is an allosteric reaction, it imposes its own limits. Remember how a tenfold change in dose produced the same graph? That's because if you throw a bunch of BPG at a red blood cell and that cell's carrying capacity for oxygen decreases, it only decreases so much. As a natural molecule, it's designed (if you want to use that term) to work within a range.

After some measurements of the dog's oxygen utilization and heart rate to confirm that it truly is in a hypometabolic state, as Cheng promised, the dog wakes up, just as the mice did. No harm done. In the last scene, we see him frolicking on linoleum without a care in the world. Cheng tells me that he was later adopted.

One made-in-China video hardly proves that AMP is safe and effective. And we're a long way from clinical trials in people. But still, I have to wonder what might be possible someday.

Indeed, it's difficult to watch that video clip of the badass beagle and not think about how some version of AMP might be used, someday, on a cardiac arrest victim. Paramedics arrive at the scene, apply cooling blankets, and give an injection of AMP. Then, with a dramatically lowered metabolic rate, and concomitantly lowered oxygen requirements, the patient is taken to a hospital where he can be evaluated and resuscitated in a controlled way. We'd treat a cardiac arrest in much the same way that we'd treat a broken leg—as an urgent problem certainly, but one that can be managed carefully and methodically, rather than with a pell-mell ambulance ride through city streets.

STUART LITTLE'S NEXT ADVENTURE

"Ah, there he is."

I ask Cheng if he's sure. That's my mouse? He looks like he's starting to wake up from a very deep sleep. Cheng rolls him onto his back with a gloved finger, and #0011 rights himself. He paws at the wood shavings in front of him. I swear he yawns.

Cheng is grinning. "See?"

He points out that the mouse would not have attempted to right itself an hour ago. Back then, he was a small furry pillow. But now he's starting to wake up.

Together we watch as little #0011 wakes up and warms up. His temperature gradually swings back to normal. By the time Cheng and I leave the lab, the little guy is scampering around his cage. Not even a hangover.

There are limits to these AMP injections. According to Cheng, the mice wake up after five to seven hours. And if you keep dosing them with AMP, they die. So Stuart Little's next adventure will probably not be to Mars in a suspended animation pod.

Cheng explains that the challenge is glucose. If AMP slows the breakdown of glucose into energy, then there's a limit to how long an animal can survive. If you don't give those cells another form of energy, they're going to die.

I think for a moment about what I know about metabolism. What about fat? I ask.

Cheng nods enthusiastically. In a hypometabolic state that lasts for a few hours, he explains, the main source of energy is glucose. But in hibernation, it's fat. The trick is going to be to convince cells to switch over. That's a challenge, but it's an opportunity, too. If you can induce that switch, that opens up a whole new door.

We walk down the tiled hallway toward the elevator, and although our visit is now officially over, there's still a whole list of questions I'd like to ask. But none of them has to do with BPG or about what goes on inside cells. Instead, I find myself thinking about the implications of this line of research.

In the elevator, I push Cheng to think about what might be possible for people. If you can put a mouse into a state of hypometabolism with a simple injection, what might that mean for accident victims? Or soldiers with battlefield injuries?

Cheng is reluctant to speculate. When we'd first met, he warned me that his work focused on mechanisms of hypometabolism, not on clinical applications. And throughout our conversation today, he's been reluctant to take more than a few steps beyond the comfortable zone of mice and into the domain of men.

But Cheng may not have a choice. This is the world, remember, that turned Mitsutaka Uchikoshi into an international media sensation. And it's the world that has become convinced that, no matter what its clinical applications, suspended animation is a great tool for interstellar travel.

The general public is likely to find results like Cheng's irresistible because the potential here is very hard to ignore. This entire line of research, and Cheng's in particular, has an elegance and clarity that seems to demand that we think of implications and applications. What started as a harmless—except to squirrels—fascination with hibernation is gradually turning into a clinical science that could save lives.

Cheng reminds me of the example of gene therapy. Back in the '80s, gene therapy was the next big thing. We thought we were going to cure genetic disorders and cancer. But, Cheng argues, we jumped into clinical trials too soon, before we understood the mechanisms, and as a result nothing happened.

He contrasts that with science's approach to HIV. First we worked out the mechanisms, he says, and the biology. Then we created and tested treatments. Now, finally, we're getting to vaccines. And that approach worked, he points out. People are living for decades with HIV, whereas gene therapy hasn't cured genetic disorders and cancer yet.

As Cheng walks me to the street to find a taxi, we talk about how gene-therapy research was driven by pressure from investors and the public. Everyone wants to see how an experiment today could save lives next week. That's troubling to him, but it's part of the game. He shrugs.

"Human beings are impatient," Cheng says. Indeed we are.

5

The Deep-Freeze Future: Cryonauts Venture to the Frontiers of Immortality

THE THING

Picture this: It's a chilly late afternoon in March, and you're walking alone through a deep, dark northern forest. The past few days have been warm and sunny, but there are still pockets of dirty snow under tree trunks and around boulders. Everywhere you look there's an unbroken carpet of last fall's leaves covering the half-frozen mud beneath. You notice that with every step it's an effort to pull your heavy boots free. You're getting tired.

Night is coming on quickly, and the temperature has dropped twenty degrees in the past hour. The wind is working its way inexorably through your jacket and you keep thinking that the road you're looking for should have appeared a long time ago. There's probably only an hour of daylight left, and you're trying to avoid the obvious conclusion that you're lost.

Then you hear a noise. It sounds like leaves rustling. But you spin around and there's nothing there.

There's the sound again. It's even closer now. It's coming from right in front of you. Again, though, there's nothing there. Your heart begins to race, and you start to hyperventilate. You tell yourself there's nothing to be afraid of, but you are, undeniably, very afraid.

Then you see it, right in front of you. It came from nowhere and it's looking right at you. It's got long webbed toes that end in clawlike fingers, which are reaching toward you. It opens a wide, horrific gash of a mouth. You jump back, heart pounding, as the perspiration beading on your forehead turns to ice.

Can you outrun it? And would a girlish scream be justifiable under the circumstances?

In brief: yes, and no. Specifically, a scream would be a little out of line, unless you want to be the laughingstock of the forest.

Because no matter how much this story resembles a scene from a low-budget remake of John Carpenter's *The Thing*, the imminent danger is really negligible.

The creature that just clawed its way out of the ground in front of you is not a threat. Honestly? It's kind of cute.

I can report this with 100 percent certainty because I'm looking at one now, in the Philadelphia Zoo. It's about three inches long, is a murky brownish-green color, and has bulbous, trusting eyes. It's an American wood frog (*Rana sylvatica*), and it's perfectly harmless. I think this one just winked at me.

THE FREEZING TRICK

Like many amphibians, the wood frog survives harsh winters by burying itself under loose dirt and leaves and snow. Just a few inches below the surface, it can survive even the worst weather. But what's remarkable about this little guy is not that it can survive a harsh northern winter but *how* it manages that trick.

What's its secret? Unlike most northern animals that hibernate, the wood frog doesn't simply slow its metabolism. Instead, it allows itself to freeze.

In fact, it can let its body temperature drop to a few degrees below freezing (-2 degrees Celsius). At that point, it stops breathing and its heart stops beating. It's not just dormant, it's dead. And it can stay that way for up to several weeks until it's jolted back to life.

Now, I'll admit that allowing yourself to freeze doesn't seem like much of an accomplishment. But this little frog has pulled off a trick that a hundred years of science hasn't managed to replicate on any other animal. Freezing is extraordinarily difficult to survive because it does three very nasty things to our cells.

First, freezing creates ice. That's hardly a surprise, but it's a serious problem because when water (of which our bodies are primarily made) turns to a solid, it forms jagged crystals with lots of uncomfortable pointy, spiky ends. So at a microscopic level—the level at which our cells are living happily—ice crystals become lethal threats to a cell's membrane, poking ragged holes and tearing them apart.

Second, as if that weren't scary enough for a small cell, remember that as water solidifies, it expands. One gram of ice at 0 degrees Celsius has a volume that is about 9 percent larger than one gram of water did at 1 degree Celsius. So when water outside a cell turns to ice, it presses in on the cell, squeezing it in a way that one can only imagine is very, very uncomfortable.

And there's water inside a cell too (at least 80 percent). When a cell freezes, water gets pushed out. That's when things start to get very unpleasant.

Have you ever put a can of beer in the freezer just for a few minutes to chill it quickly? And have you ever gotten distracted enough that the "just for a few minutes" part didn't work out? How long did it take you to mop beer off those bags of frozen peas? You can imagine how our cells might get a little nervous about all this talk of freezing.

Third, there's the problem of what happens to all of the electrolytes that float around in solution, inside and especially outside of cells. For instance, as ice forms in the fluid that surrounds a cell, there's no room in that crystal structure for ions like sodium, chloride, calcium, and potassium. So as water freezes, those ions build up in the remaining liquid, in increasingly high concentrations. Those high concentrations, in turn, pull water out of cells by osmosis, causing the cells to shrink to the point at which their membranes collapse irreversibly. This results in cell death, which is generally considered to be a very bad thing.

In short, our cells really, really don't like to freeze. They're not made for it, and they're not good at it. And when freezing happens despite their objections, they die. Yet somehow wood frogs have managed to sidestep these perils, and *their* cells freeze and thaw happily.

Humans, on the other hand, haven't managed to come close. Even with the benefits of a hundred years of science, the best we've been able to accomplish is freezing small parts of animals and people. And I do mean small. One of the first reports of successful cryopreservation was of chicken sperm in 1949. More recent advances have led to the successful freezing and thawing of human sperm, pancreas cells, red blood cells, corneas, and heart valves. But that's about it. Even now, scientists haven't managed to achieve what Kermit and his pals are able to accomplish every winter.

Nevertheless, you can't help but wonder whether freezing might be possible eventually for mammals. And large mammals, like people.

That would be a neat trick. Think of everything you'd have to look forward to if you could step into a freezer and then out again in, say, a thousand years. But is it possible?

Unfortunately, this is not something one can ask respectable scientists. There's something about the topic of freezing people that seems to activate some sort of allergic reaction. No, if I hope to find out whether freezing people might be possible someday, I'll have to go someplace where respectable scientists are scarce.

TIME FLIES WHEN YOU'RE DEAD

The man sitting next to me in this packed hotel conference room is in his thirties, with tousled brown hair, blocky glasses, and a thin goatee. He looks like an aging hipster playing hooky from family life for the weekend. He is, as near as I can tell, an ordinary guy. That impression is confirmed as he turns to me and introduces himself.

"Hi! I'm John. I'm just an ordinary guy," he says cheerfully, nodding vigorously as if to reassure me. "Just an ordinary guy," he says again. "I'm not a paid-up member yet. You know. In the program. In fact, my wife doesn't even know I'm here."

"The program" is the reason why the three hundred people around me have packed themselves into this auditorium at a resort in Scottsdale, Arizona. All of them are here at this cryonics conference to learn about the latest advances in the science of freezing people and then (hopefully) reanimating them. Many are here because they've already agreed to invest the $200,000 that will buy them a membership in the cryopreservation club when they die, and they want to know what they can expect. But a skeptical few, like John, are still weighing their options.

In the course of the morning's lectures, I've discovered that John and his fellow attendees are a mixed bunch. There are the "ordinary guys," but they seem to be few and far between. Much more common are what I've come to recognize as the "true believers," who seem to be motivated by a mix of enthusiasm and curiosity that is at the same time baffling, admirable, and just a little bit scary. You can recognize them because they're the ones asking arcane questions ("Wouldn't preservation with potassium chloride increase the likelihood of successful cardiac resuscitation?").

This bunch describes themselves, first and foremost, as "cryonauts." They also employ the terms "futurist" and "transhumanist" and "strategic philosopher."

Many are inspired. Some are visionary. Others, like John, are ordinary. More than a few, though, seem to be utterly and completely bonkers.

To spend a day in the company of cryonauts is to oscillate between astonished enthusiasm and incredulous skepticism in a cycle that is punctuated by sudden and uncontrollable urges to laugh hysterically. And yet, if you can put all that aside for a moment, you have to wonder whether, just maybe, they're onto something here.

That's what I'm here to try to find out. I've come to the fortieth anniversary conference of the Alcor Life Extension Foundation, whose Web site defines cryonics as "the science of using ultra-cold temperature to preserve human life with the intent of restoring good health when technology becomes available to do so." It's a nonprofit organization that promises to freeze the bodies of recently deceased cryonauts, safeguarding them at a chilly -196 degrees Celsius until such time when medical science can reanimate them and can cure whatever diseases ended their lives in the first place.

I turn to John and tell him that I'm new to all this. I admit—hoping this will draw him out—that I'm just a little bit skeptical.

John's mouth tightens and he adjusts his glasses nervously. "I don't really like talking to skeptics," he says carefully.

Oh boy. This, I'm thinking, is not going to go well.

After a tense pause, though, and perhaps against his better judgment, John relents.

"Well," he says, "here's the thing. You know how much medical science has advanced in the past hundred years, right?"

I nod.

"So it stands to reason that there are people dying every day of diseases that, a hundred years from now, won't be much more serious than the common cold."

John looks at me expectantly, as if he's hoping for agreement and maybe validation. But I'm not nearly as convinced of this second point as I was of the first. For instance, the common cold has survived happily for the past thousand years or more, untroubled by the scientific advances that have produced open heart surgery, antibiotics, and breast implants. But mindful of John's earlier warning about skeptics, I offer a hesitant nod, which seems to reassure him.

"That's the theory," he says, growing more animated. "It's like putting people to sleep and then waking them up"—he pauses—"when it's safe." He grins. "It's like a . . . time machine, almost."

A *time machine*? I'm not sure what my face reveals as John makes this analogy, but it probably doesn't indicate wholehearted acceptance. John's grin dims a little, but just a little. Indeed, his enthusiasm seems unshakeable.

"I mean, not really, you know?" he says cheerfully. "You can't really put people to sleep for a thousand years." He waves at the roomful of people around us, laughing a little too heartily. "We all know that."

Actually, I wouldn't count on that. I'm thinking back to the earlier sessions that morning and to the questions that I heard being asked. One earnest woman, for instance, wanted to know whether it's easier to freeze a cat or a small dog. Another man suggested developing an iPhone application that could keep an eye on his stock portfolio while he was frozen. So I'm not ready to take for granted this audience's understanding of anything. But John, at least, seems to have retained his grip on reality.

"Sleeping? Hibernation?" He laughs. "That's just flaky science. But cryonics? That's real."

I nod enthusiastically, as if he's convinced me. As if we're on the same page. Brothers on ice.

But I'm curious. If cryonics is real science, why isn't he a member? So I ask him whether he thinks that kind of reanimation is really going to be possible for people like him. For people like us, I correct myself.

His grin fades, but then he nods seriously. "I do," he says finally. "I really do. Of course, some of this is hype," he says in a half whisper, waving at the stage where the next presenter is getting ready. "It's a business, right? They're trying to sell a service, so of course they're going to exaggerate what it can do. But that's OK. I'm in IT sales—my job is to get companies to buy bundled telecommunications packages. So I've got no problem with a sales pitch as long as the product makes sense. Nothing wrong with that."

But what does he think the chances are of being reanimated in a thousand years? A hundred percent? Fifty percent? One percent?

John thinks about that for a second, then he shrugs. "One percent seems too low," he says hopefully, "but I'll admit the chances really aren't very good. And there will be lots of mistakes, probably. You don't want to be the first to come out of the deep freeze, do you?"

Before I can ask him about that, the next speaker has walked onto the stage. We're about to learn about what happens to a cryonaut after he dies. I can hardly wait.

THE STRANGE AND AMAZING VOYAGE OF A CRYONAUT

What little I know about cryonics sounds particularly unsavory, and for this part of the program I'm fully prepared for a titillating exposé, full of late night skullduggery and cryopreservations carried out in garages and basements.

But it turns out that I couldn't have been more wrong.

The slim, petite woman with a pixie haircut who has just taken the stage is Catherine Baldwin, the general manager of Suspended Animation, Inc. (SA), a Florida-based company that has focused on advancing the science of body retrieval and preservation. Although this is an Alcor conference, SA is prominently featured because Alcor's success in preserving someone depends on how quickly a cryonaut is transported to the Alcor facilities, and how cold they are when they arrive. SA has five employees and numerous consultants, but Baldwin is its most visible face. As she talks, I can begin to see why.

Carefully groomed and elegantly suited, she delivers her spiel rapidly but flawlessly. She talks forcefully, with a calibrated cadence that is careful, polished, and a masterpiece of rhetoric. But she speaks in architecturally complex paragraphs that often carry clauses nested within clauses. In short, she looks like a CEO and talks like a scientist. As she introduces herself, it turns out she is both. Before her role at SA, she has held a variety of positions, including as a biologist at UCLA. She's pulled together that experience and those roles into a presentation that's as polished as any medical grand-rounds lecture I've ever seen.

Using slides that flash rapid-fire on the screen, Baldwin whips a rapt audience through the cryopreservation process. That process begins, she says, when a patient has a health crisis, such as hospitalization for something serious like heart failure. At the earliest hint of trouble SA sends out a team, staffed from lists of dozens of physicians, nurses, EMTs, and medical technicians around the country.

She notes that these are highly skilled professionals, and that they undergo further training for the cryopreservation work that they will be asked to do. For instance, she mentions that all of SA's field teams do a stint at a US Department of Agriculture facility. There they practice freezing and perfusion on what she describes as a "human-sized animal cadaver."

Now, I'm thinking very, very hard, but I can't think of anything that is both human-size and . . . well . . . not human. All I can think of is a six-foot, 180-pound wood frog. But that thought is terrifying.

Sometimes the SA team is mobilized and their gear is packed onto a plane, or one of two trucks, if they're within driving distance. And soon—ideally within six hours—the team arrives with their gear. That gear is surprisingly technical, and would be a credit to most hospitals. For instance, there are ice baths for rapid cooling of the body, and medications like heparin to prevent blood clotting. And there is a portable cardiopulmonary bypass machine, which is managed by the bypass technician and the cardiothoracic surgeon.

A cardiothoracic surgeon? This is hardly the basement science experiment that I was expecting.

The team goes to work and the surgeon inserts two large intravenous cannulas, one each in an artery and a vein. Then the bypass technician connects them to the bypass machine, allowing the patient's blood to be flushed out rapidly and replaced with chilled organ preservative fluid. Then they try to cool the patient to as close to 0 degrees Celsius as possible.

As I'm thinking about this, my attention is dragged back to the stage by a point that Baldwin is making very earnestly. Their goal, she says, is fifteen minutes from death to bypass. I'm thinking that this is a very optimistic goal, because what often happens to cryonics patients is nothing close to that.

Consider the sad case of Alcor patient 113. John Monts is an example of what happens when things don't go as planned, and when you don't have someone like Baldwin on your side. Monts died October 31, 2012, at the age of sixty-eight. He died alone in a hotel room, and a pathologist's report estimates the time of death at approximately eleven p.m. So an entire night passed, and part of the next morning, before his body was found. Then the suspicious nature of his death made it a coroner's case.

Delays ensued. In fact, it wasn't until November 4 or thereabouts that Aaron Drake, Alcor's medical response director, performed the neuroseparation. That is, he cut off Monts's head and plunked it into a bucket of dry ice. (Yes, Alcor freezes brains in a process that is just about as gruesome as it sounds. But more about this later.) Finally, on November 7, John Monts's brain began its final descent to the desired temperature of -196 degrees Celsius in liquid nitrogen.

I'm betting that if John Monts was expecting a fifteen-minute drop to 0 degrees, he would have been dismayed by this pace of events. That lonely evening spent at hotel-room temperature probably wasn't part of his plans, for instance. Nor, probably, was he counting on a couple of days spent in a morgue's cooler, at roughly the temperature you would use to keep a head of lettuce from wilting. So I'm thinking that although Baldwin's slick efficiency is impressive, and although she might be able to freeze cryonauts in a matter of hours, the reality for many Alcor members is probably much more messy.

Then Baldwin slips in something that captures the attention of everyone around me. It's her pièce de résistance: liquid ventilation. She acknowledges that it's still in the early phases of testing, but it has an undeniable geek factor, and the audience is riveted.

They're starting to use perfluorocarbon, she explains, a liquid hydrocarbon that has the potential to carry oxygen and CO_2, making it a nifty blood substitute. (We encountered this substance, incidentally, in the RhinoChill in chapter 3.) But it can be chilled far below the point at which blood would solidify, and it doesn't coagulate the way blood would if it's not flowing. So not only does it allow rapid cooling, it also maintains gas exchange (at least in theory).

Listening to Baldwin's technical description, I'm having three thoughts more or less simultaneously. The first, of course, is that this is really, really neat. The second, though, is that what she's describing sounds like wild science fiction.

But it's the third reaction that surprises me: this sounds familiar. A quick Google search on my iPhone provides the obvious explanation. Liquid ventilation is a common device used in science-fiction films including *The Abyss*, *Mission to Mars*, and *Event Horizon*.

These echoes of science fiction are perhaps inevitable. The grandfather of cryonics, Robert Ettinger, first proposed the idea in—what else?—a science-fiction short story. When he was recovering from war injuries in 1947, Ettinger learned that a French scientist, Jean Rostand, was experimenting with cryopreservation. Inspired by those rumors, in 1948 he published a short story in which freezing was the main plot device: "The Penultimate Trump."

Indeed, it's difficult to navigate the worlds of fantasy or science fiction without bumping into fanciful examples of cryopreservation. It's almost as common as its more popular cousin, suspended animation, which is easier to use as a plot device because people in suspended animation remain alive. Examples of cryopreservation include, of course, the Thing, in John Carpenter's film of the same name, which wakes up very hungry and generally unpleasant after ten thousand years frozen in Antarctic ice. And there is Captain America (arctic plane crash), and Austin Powers (frozen in space) . . . the list of fictional references to freezing is a long one.

That list of fictional cryopreservation examples is also a burdensome one if you're trying as hard as Baldwin is to convince people that this is real science, damn it. In fact, I get the sense that Baldwin is very much aware of this history and the semantic connections that cryonics has for many of us. Sprinkled throughout her lecture are attempts to normalize and legitimize SA's techniques. Again and again, she compares what SA does with medical science procedures with which we're all familiar.

I'm beginning to realize that this is the main thrust of her sales pitch. Except for a few neat tricks like liquid ventilation, she's careful to say that SA is only doing what thousands of surgeons and hospitals have been doing for years. Open heart surgery, cardiac bypass perfusion, and mechanical ventilation are procedures, she reassures us, that millions of patients have undergone. "If these techniques are safe and effective for us when we're alive," she asks, "don't you have to trust that they'll work as part of an effective cryopreservation strategy?"

That argument is highlighted by the way that she drops names and credentials. For instance, she names two perfusion companies (PEC Perfusion Resources and Perfusion.com), which she says are among the biggest and best in the field. And she happily rattles off the names of four cardiothoracic surgeons she works with, naming their affiliations with prominent academic medical centers such as Duke University and the Texas Heart Institute. She also mentions a home-care and hospice agency in California that SA partners with when a patient dies at home.

The message is clear: We may not be part of mainstream medicine yet, she seems to be saying, but we should be. And we will be.

I tear myself away from this procession of slides and look around me. Up and down my row, to my left and right, there are nine people all gazing intently at the screen. Two are taking notes, one is looking pensive, and the others are staring with slack-jawed admiration at Catherine Baldwin and her routine.

It's obvious that they're hooked. She's giving them what they need. She's offering them reassurance, first of all, that her company will be there for them. But she's also offering results. She can report time to cooling, time to bypass, and time to cryopreservation. All real numbers. Of course, there's no information about what happens next, or whether there is any prayer of waking up a thousand years from now, but there doesn't need to be. That's the point. She can take them this far, she is saying. That's her claim and her promise.

Her talk ends on a high note, vowing that she will continue to advance the field. It's a masterful ending, which neatly wraps up science, inspiration, and the reassurance of professionalism. The crowd goes wild. Even I'm ready to sign up.

But I'm still thinking about Baldwin's interactions with what must be a very skeptical health-care system, and I'm curious to learn how she navigates those relationships. Not surprisingly, there's quite a crowd around her after the lecture and so I wait my turn patiently. Finally, up close, she's warm and engaging. As I watch her talking with the woman in front of me, I see that Baldwin has the quiet competence of a surgeon. I'm thinking that she's ideally suited to sell just about anything as she turns to me with a friendly smile.

I decide not to mention the fact that I'm a doctor, and completely avoid the fact that I'm rather skeptical about the whole process. Instead I play the ingenue. What kind of reception can I expect from hospitals? Are they supportive?

She sighs. "They're afraid. You [cryonauts] do not fit into their box." And, she admits, there's the stigma of cryonics. It's still not something that people want to be associated with.

She admits that it can also be difficult to find health-care providers who are willing to do this work. She doesn't even publicize the names of the health-care providers she works with. She's found that any association with cryonics is viewed in many circles—particularly among doctors—as career suicide.

Everyone in the circle around Baldwin laughs. Including me. Then I remember that I'm a doctor standing in a roomful of cryonauts.

But Baldwin didn't get to be where she is by whining. I'm waiting for a solution, and for her sales pitch. I'm not disappointed.

She smiles. "It's up to you," she suggests. "You can choose the hospitals that will be the most supportive, and which would give you the best chance."

Her message is perfectly calibrated to the independent, almost libertarian leanings of this group: Use consumer pressure to shift hospitals' opinions. Make them more open to cryonics. Use your leverage as a consumer.

"Talk to your doctors," she suggests. "They're more open, and they know you. And," she adds, "they're the ones who can make things happen. That's the only way we're going to develop the relationships we need to make sure that we can work together to provide the best possible care."

By "relationships," Baldwin clarifies, she means the sorts of arrangements that would allow SA staff to use hospital facilities. Annoyed and perhaps a little embarrassed by the necessity of parking an anonymous-looking van at the hospital's loading dock, she wants a seat inside the tent. She wants to be able to use hospital ORs the way that transplant surgeons use them to harvest organs.

"And why not?"

I'm hoping that's a rhetorical question, because I can think of a whole range of reasons why not. There is, of course, the rather obvious observation that they're proposing, in all seriousness, freezing people. And reviving them in a thousand years. That is enough to get you kicked out of most hospital executive suites.

Then, too, there is the language I've been noticing as I've talked with many of those around me. This group talks about people who are irreversibly dead as opposed to those who are planning on making a second trip through life. Old cryonauts don't die, they "deanimate."

So as I thank Baldwin and she turns to the next person in line, I'm thinking that cryonics' road to legitimacy is going to be a long, difficult one. But whether she's able to develop those relationships she wants so badly will depend not on what Baldwin can promise but on what happens after she's done. Whether we take her team seriously depends almost entirely on whether there is, in fact, a "science" of cryonics that will take over after Baldwin and her team do their thing. And that's next.

TODAY A RABBIT, TOMORROW A CRYONAUT: THE NEW SCIENCE OF VITRIFICATION

Baldwin's enthusiasm was inspiring, and the science she described was impressive. But the work that her teams do in cooling and transporting is only effective if someone else can fully freeze a patient safely. And by "safely" I mean in a way that makes it possible that they'll be brought back to life someday.

But how?

We're about to hear the answer from a pharmacologist named Greg Fahy. He's tall and intense, with an oversize mustache, and he talks like an aging hippie scientist. But he also has the calculating, logical speech of someone who once thought very, very hard about becoming an accountant, and who decided in the end that he wanted something more intellectually challenging. And he seems to have found it.

He warns us that cryonics is technically very difficult. And the slides flying by in his presentation suggest that there are multiple pitfalls waiting to claim the unwary cryonaut. These slides sport cheery headings like "Approaches to Preventing Brain Shrinkage" and "Preventing Brain Blowout."

But at the heart of the cryonics puzzle, Fahy says, is the physics of freezing and thawing. That's what it's all about, and that's where cryonics must succeed. Fahy reminds us, with an intensity that is perhaps more appropriate for communicating the world's total financial collapse or the untimely death of your dog, that the problem is—he frowns for emphasis—ice.

So how do we get ourselves down to very low temperatures without letting the water inside us freeze?

What Fahy is talking about next might hold the answer to that question. Vitrification, he says, is the process by which each one of us could—and perhaps someday will—be cryopreserved. Literally "to transform into glass," vitrification refers to the process by which a soft, squishy, pillowlike frog, or a human being, is turned into a solid with the consistency of your kitchen counter. As it's used in the cryonics world, though, vitrification refers to the process by which that cooling and hardening takes place without ice formation. It's no mean feat and, apparently, the journey from squishy to solid, without encountering ice along the way, is an exceedingly difficult one.

The first step along that road is to find a way to inhibit ice from forming. Usually that's accomplished by replacing water with something else that won't develop ice's jagged, knifelike crystals and that won't create dramatic swings in electrolyte concentrations. You'd also want something that won't expand as it cools, causing you to burst at the seams like something out of a bad horror film. The answer, it turns out, lies in cryoprotectants—substances that prevent ice formation and allow vitrification to occur without damaging cells.

Remember the humble wood frog? Cryoprotectants are its elegantly simple secret. As the weather gets colder, the frog stockpiles natural compounds like glucose that act as natural antifreeze.

What is even more interesting is that this protection seems to be triggered by ice formation. That is, as a frog starts to get cold, tiny ice crystals begin to form. When this happens, frogs respond by depleting glycogen to make glucose, effectively constraining that ice formation. Then, rather than becoming frozen, with all of the attendant problems of ice, the frog vitrifies, turning into an ice-free little piece of statuary. And the best part is that glucose and other cryoprotectants are all natural and frog-made, so when it warms up, the frog simply metabolizes them.

But what about the rest of us who don't happen to have the ability to generate our own cryoprotectants? Well, the preservative solutions used for freezing various parts of people, such as corneas, rely heavily on glycerol. Glycerol is a by-product of the breakdown of fats, which usually appears in soaps and pharmaceuticals as a thick, viscous liquid. It acts as a protectant, elbowing water out of the way and taking its place.

Organ preservation also uses dimethyl sulfoxide (DMSO), a spectacularly efficient solvent that crosses cell membranes easily and drags anything dissolved in it along for the ride. I remember working with it at a part-time job in a chemistry lab in medical school. You always knew if you spilled some on bare skin, because it would migrate amazingly fast from there into the body. Within a minute after a few drops landed on a finger, for instance, you could taste it. (In case you're curious, the taste is not entirely unpleasant and reminds me of day-old garlic bread. But don't try this at home.)

However, glycerol and DMSO just don't work well enough when you're trying to freeze a whole organ, or an animal (or a person). You still get ice and all of the problems that go along with it. So the cryonics industry has turned to ethylene glycol, the chief ingredient in antifreeze. You know, the green goo in your car's radiator.

The problem, though, is that ethylene glycol is highly toxic. If you drink it, it gets metabolized into glycoaldehyde, and then to glycolic acid. Glycolic acid, in turn, has a tendency to bind to calcium, forming calcium oxalate crystals. In tiny amounts, these crystals don't cause much harm. But in the amounts that are typically used in cryonics, one can only imagine that a cryonaut would wake up with organs that have turned to stone, not to mention a wicked hangover.

Even with antifreeze, we've learned through trial and error in preserving small human organs like heart valves and corneas that the vitrification protocol is very dicey. The cooling and thawing processes have to be rapid and homogenous. You need to get the entire frog—or person—down to the desired temperature as quickly as possible. If cooling is too slow, there will be time for pockets of cryoprotectant-free fluid to freeze into ice. For the same reason, the cooling has to be even. That is, the core of a frog has to freeze at the same rate that its little webbed fingers do.

So when cryopreservation works, it works on small things—blood cells and corneas and embryos and heart valves. It's easier to freeze small things quickly and evenly. And you don't need to use compounds that are probably best left in your car's radiator. But scaling up is very difficult. In fact, science has yet to adequately preserve anything much bigger than an acorn, and we haven't come close to replicating the wood frog's achievement.

However, there have been a few successes that provide grounds for cautious optimism. In 2005, for instance, a team of researchers was able to extract rat hearts and vitrify them. As cryoprotectants, they used proteins derived from arctic fish (notably the oddly named ocean pout, *Macrozoarces americanus*), which are accustomed to subfreezing temperatures. These so-called antifreeze proteins allowed the researchers to preserve the hearts and to transplant them into the peritoneal cavities of another set of rats. In their new homes, the hearts (unconnected to a circulation) began to contract. The same team reported later that in a subsequent experiment they were able to produce normal electrical impulses and conduction in the revived hearts.

Fahy mentions a study of his own that was even more ambitious. A kidney was removed from a rabbit, vitrified, thawed, and then reimplanted. One can only imagine what the rabbit thought about this procedure, which, from its perspective, probably seemed rather unnecessary. The rabbit, Fahy reports happily, lived. So score one for cryonics.

Later, I looked up Fahy's published article and discovered that the experiment he described was carried out on two rabbits, one of which died. The lucky rabbit who lived only did so for nine days. I suppose in rabbit years, that's probably two human months. To be fair, Fahy wasn't being disingenuous. Those nine days were days that rabbit really shouldn't have expected.

And the applause that greets the end of Fahy's lecture suggests that the audience agrees that those nine days are a victory. Today a kidney, tomorrow a cryonaut.

STRUCTURE, FUNCTION, AND THE FINE ART OF TAXIDERMY

As the applause for Fahy's rabbit trick begins to subside, I elbow my way to the front of the room to ask about the wisdom of using antifreeze in people. The line in front of Fahy is surprisingly short. There is just one middle-aged man ahead of me and I catch the tail end of his question: "Aren't you biased toward chemopreservation rather than cryopreservation just because current techniques of connectome mapping require fixation?"

Translated into English, this man's question hints at what's been nagging me. Basically, my fellow attendee is suggesting that the common metric for evaluating whether cryopreservation is successful—the preservation of structures as viewed under a microscope—doesn't take into account what kind of damage cryopreservation does to function. With the help of compounds like ethylene glycol, you can preserve tissue and make it look nice at 1000x magnification. But given how toxic these compounds are, there seems to be a high probability that what you'll be left with is just a shell.

I'm thinking that this emphasis on structure, rather than function, makes cryonics a little like taxidermy. The result looks nice. And an expertly stuffed lion is probably indistinguishable—at a safe distance, at least—from a living one. But that stuffed lion is never going to roar again.

Fahy's answer to the man's question is vague. Really, who knows? But he reminds us that cryoprotectants and vitrification have been used in other settings to preserve body parts. He mentions that lone rabbit survivor too. I'm thinking there's a lot riding on nine days in the life of a small, probably very confused, lagomorph.

Finally, Fahy turns to me and I ask him a question that I think I already know the answer to. Kidneys, I suggest, are really pretty simple. They're just tubes with membranes that act as filters. They're simple enough that we've been using artificial kidneys (dialysis machines) for decades. But what about the brain? What are the chances that all of those tangled neural structures could survive the vagaries of cryoprotection and freezing?

What would cryoperfusion do to a whole brain's function?

He ponders that question for what is, for me, an excruciatingly long time. "The answer," he says finally, "is that we don't have a clue."

COULD CRYOPRESERVATION WORK? MAYBE? EVENTUALLY?

Fahy's reluctant summary is emblematic of this entire field. No one knows what works, or what might work. Worse, no one knows whether the things they're doing to people right now are working. That point is about to become very, very clear.

During the break, I'm sipping coffee and trying to eavesdrop inconspicuously on the conversations around me. But my attention is pulled inexorably to the screen at the front of the auditorium that displays a never-ending parade of CT scan images. Shown in vivid shades of blue, green, orange, and yellow are brains in finely detailed sections, animated to show various levels in all their pixelated wonder.

Complemented by the soundtrack soft elevator music in the background, they're like something a person might hallucinate on acid. Especially if that person were, say, Oliver Sacks. They're mesmerizing, even beautiful.

A question begins to form in my own dull, gray brain. Where do these colorful brains come from?

Fortunately there's someone standing quietly up against a nearby wall who might be able to help me, I think. Ben Best is the director of research oversight at Alcor, which has contributed these images. Best looks like someone whom an interloper like me can ask even the silliest questions. Wild wisps of gray hair fling themselves outward in a halo that surrounds an otherwise bald head, and his thick glasses, oversize ears, and a chunky tweed sport coat make him look like a gentle English professor.

I introduce myself and ask my question. These brains—where do they come from?

"They're the brains of cryonauts," he says simply.

I look around the room, envisioning the bunch of five or six attendees nearby getting together over wine and cheese for a CT party. Like a Tupperware party for transhumanists. I ask him how many of the people here have been scanned, and he shakes his head.

"No, patients in cryostasis." He's grinning proudly now, and points at the current slide excitedly. We're looking at the cross-section of a head, cut about the level of the eyebrows. The entire section is a bright, happy orange. "Look," he explains, "that's all cryopreservative."

So does that count as a successful preservation? Did it work? It turns out that how you answer this question depends on what you mean by "successful." I'm about to hear Best's definition.

"Absolutely," he says. "Of course, we'd like to get rid of all of the ice." He points to a corner that looks like it's at the top of the right temporal lobe. "Except here, that's a little bit of ice. But not much," he says hopefully. "Probably less than one percent."

But . . . how does he know that the preservation was "successful"? I press.

"Well, there's no ice. As I said." He's starting to look a little annoyed, if it's possible for everyone's favorite English professor to look annoyed.

So that's his definition of a successful preservation? They look for ice crystals? I'm thinking that's a very, very low bar. It's a little like saying a surgery was a success because you managed to cut someone open. It's necessary, of course, but surely not sufficient.

I'm skeptical because I'd like to see a more meaningful definition of success. Ice-free preservation is a good first step, but it's just a first step. I'd like to see that frozen organs can be thawed and shown to function normally. But with the exception of one rabbit kidney, it seems like my definition of success is a long way away.

"Of course there are other things we track," Best says. "Like time to dry-ice temperature, and time to cryostasis in liquid nitrogen. We also watch for evidence of temperature swings prior to cryopreservation. But," he says, "this is hard evidence. We can look at these scans, and see how we're doing." And, it appears, he's confident that they're doing pretty well.

THE TRAVELS OF A WANDERING CRYONAUT

And maybe Best and company are doing pretty well. Now. But things haven't always been so rosy.

A case in point, which is either a sign of progress or a cautionary tale, is that of James Hiram Bedford. Bedford, it seems, moved around a lot. Not during his life, which ended January 12, 1967, due to renal cancer in a California nursing home. No, it was after he died that he began to roam.

Bedford was frozen initially by a man named Robert Nelson, who was then president of the Cryonics Society of California, under circumstances that appear to have been bizarre, even by cryonics standards. For instance, no one did anything for the first hour or two after Bedford "deanimated." Then they packed him in ice and started CPR. Then they injected him with DMSO. (I suppose it couldn't hurt.) Then they wrapped him up and put him in a box and covered him with slabs of dry ice, later plunking him into a vat of liquid nitrogen.

Bedford was moved within days to a company called Cryo-Care (now defunct), and then in April 1970 to another company called Galiso (still around but, its Web site suggests, staying as far away from freezing humans as possible). He was there for about six years until Galiso had to get rid of him, apparently because insurers had liability concerns. It's unclear what those concerns could possibly be, given that Bedford was really quite dead.

So poor Bedford was packed up and moved again. This time he was taken on July 31, 1976, to a company called Trans Time, Inc. (still in existence as of 2012). After less than a year, though, Bedford's family, perhaps tired of all of this moving, loaded him in a U-Haul trailer on

June 1, 1977, and took over his upkeep in DIY mode. That's right. They took him home.

He seems to have spent some of the next five years at a storage facility and some at home with his family. It's not clear where they put him, or how they managed to afford the biweekly deliveries of liquid nitrogen necessary for his upkeep, but one hopes that after all that wandering Bedford at least had a room of his own. Eventually, though, in 1982, his family gave him to Alcor's facilities, then run by a company called Cryovita (now defunct). From there he was moved in 1991 to Alcor's facilities in Scottsdale, where one can only imagine he'd have heaved a sigh of relief if only he could still breathe.

But he had one more indignity to endure because his last (and, one would fervently hope, his final) transfer required some unpacking. This, someone thought, was a great time to take a peek at poor Mr. Bedford. So, how well did he do?

Well, the good news is that once they cut away the container and removed him from the sleeping bag that he was stored in (the traditional Cryonics Society of California method), they found chunks of ice in cube form. This suggests that if any thawing had taken place along the way, it hadn't been severe enough, or prolonged enough, to disrupt the ice structure. It also meant that Bedford was surrounded by enough ice for quite a few martinis.

The bad news, though, is a bit more involved: "The skin on the upper thorax and neck," the report says, "appears discolored and erythematous [inflamed] from the mandible to approximately two cm. above the areolas." (This, by the way, is the point at which the squeamish will want to stop reading.)

The report continues in the same dispassionate tone: "The nares are flattened out against the face, apparently as a result of being compressed by a slab of dry ice during initial freezing. Close examination of the skin on the chest over the pectoral area disclosed sinuous features that appeared to be fractures."

That's right. Apparently the freezing process leads to cracks. You know, like an ice cube.

Finally, buried in the latter half of the report is this note: "There is frozen blood issuing from the mouth and nose."

OK, you'd think this information would make prospective cryonauts pause and think very, very hard about what they're getting themselves into. Flattened face? Cracked skin? Blood? This makes Bedford sound like more like the loser of a bar brawl than someone who is boldly voyaging forth into the future.

At the very least, this doesn't sound like a description of someone who is going to be jumping for joy when he wakes up. Nor, in all likelihood, is he going to be jumping for any reason when he wakes up. In fact—and I'm admittedly going out on a limb here—he doesn't seem likely to wake up at all.

SCIENCE AND THE FUTURE

Granted, Bedford's is a worst-case scenario. Remember, he was frozen by a friend under circumstances that sound more like a frat party hazing gone terribly wrong than any sort of medical procedure. Then he was moved around like an unwanted piece of furniture for the next twenty-five years. Actually, given all of that, Bedford's level of preservation is remarkable. Still, Best and his team have their work cut out for them if they are going to convince skeptics that they can preserve bodies in a state that will result in anything other than a reanimation scene reminiscent of *Night of the Living Dead*.

Will that ever be possible? After a day of lectures and conversations I'm still on the fence. It seems like a stretch to imagine that anyone could be frozen and revived in the way that Alcor promises. And yet, there are frogs out there who manage to do something similar. And as John the ordinary guy pointed out to me this morning, we can do all sorts of things today that no one dreamed possible ten years ago. So why not?

In hopes of a dose of scientific realism from someone I can trust to tell it like it is, I go looking for Ken Hayworth, one of the morning's speakers. Up close he's tall and impossibly thin. He looks like the geek who was ostracized in high school who has grown up to become a professor at Harvard.

I ask him, as one scientist to another, does he really believe this cryonics stuff?

"Well, I'm an Alcor member," he admits.

That surprises me.

"But I'm a skeptical Alcor member," he continues. "I get no solace from that Alcor membership. If I went into a surgical procedure with a five percent chance, I wouldn't go into that procedure comfortably."

And yet, Hayworth tells me, he wants that solace. Moreover, he thinks cryopreservation is theoretically possible. He imagines a world where cryopreservation is standardized and high quality, done in hospitals with the science and transparency of other medical procedures.

I'm still thinking about that hopeful view of the future as the first day comes to a close. I'm walking across the dark parking lot when I hear trudging footsteps behind me, then muttering. I turn around to see a man in his seventies, slightly hunched over and wrapped in an oversize sport coat despite the warm night. He catches sight of me and gives a quick bob of his bald head in acknowledgment, but he doesn't slow his pace. I have to hurry to keep up.

I ask him what he thought of the day.

"Crap. It's all crap, isn't it?"

I'm sensing that this is a rhetorical question.

"Every year I come here expecting to hear something new, but I never do."

I protest mildly that I thought I'd heard some good news. Response times are getting better, and the preservation statistics—

"Heard that last year," he interrupts me. "And the year before that. Pisses me off. You young guys, you don't care. You've got time. But us? We're getting old. We could kick off any minute."

I don't know what to say to that, but he stops and, without a word, climbs into an oversized pickup truck with a large white LIFE EXTENSION VITAMINS sign jerry-rigged onto the back. The big diesel starts up and the grumpy skeptic roars off into the night.

SCIENCE IN SEARCH OF A SAVIOR

The next morning I'm back, but now I'm pondering a different angle. Yesterday I was focused on the science of cryonics, and I was asking whether it could work. I'm no closer to an answer, yet I'm starting to suspect that this question is just too narrow.

In the end, what will make cryonics work—or not—isn't only the science behind it. That's clearly important, but there's more. What also matters, and what will determine whether you're going to wake up in a thousand years, is not just the technology itself but how (and if) it's used.

And that depends on what people think of it. The patient is a believer, presumably. But what about the family members gathered around? And doctors and hospital staff and EMTs? And funeral directors and coroners?

If any one of these important actors begins to put up roadblocks, then it doesn't matter what wizards like Hayworth and Fahy have cooked up in a lab. And it doesn't matter what sorts of portable technology Baldwin and her team have assembled. None of that science will get within a mile of the patient.

What cryonics needs, I'm starting to think, is a PR savior. And judging by the screams and cheers as the latest speaker takes the stage, it looks like they might have found one. I know this because I can already hear people yelling his name with a religious fervor normally reserved for Third World dictators and NBA forwards.

Max More, as of 2014, is the CEO of Alcor, and has become its de facto public face. And since Alcor is the biggest, most successful player in the cryonics game, More is also the de facto spokesperson for the entire field. Born Mark O'Connor in Bristol, England, More comes to the savior role armed with rock-star good looks and a smooth British accent. With close-cropped blond hair, a thin goatee, and a predilection for silky sport shirts unbuttoned to mid-chest, he looks less like a scientist than a superstar yoga instructor or a celebrity chef.

In short, More is cryonics' best hope for reaching the mainstream. And the cryonauts here seem to recognize that.

In fact, PR seems to be More's primary focus these days. His driving effort is to try to build up the legitimacy of cryonics as a science, as an industry, and as a movement. Throughout the conference, and in conversations during intermissions, he carries with him a seasoned politician's ability to stay constantly, persistently, and unnervingly on-message.

That message started with some of his opening remarks, in which he warned us all that Alcor membership is just one step on the long road to immortality. "Once you're a member, it's not *good-bye, death*," he warns. "Absolutely not. Nothing could be further from the truth."

He goes on to paint a gruesome picture of all the things that can still go wrong. Your family may ignore your wishes. EMTs might not continue CPR once you're dead. Hospitals might not permit recovery teams into the ICU. Coroners could insist on an autopsy. Each of these examples incites another wave of what is becoming a crescendo of grumbling in the audience.

"But," he adds, "there are lots of things you can do to improve your chances." He treats us to a dramatic pause that puts the audience on the edge of their seats.

First, he suggests, the cryonics world needs to develop communities. Cryonauts need to watch out for one another. Protect one another. Make sure that our wishes are honored.

Second, advocate for yourself. Make a video for friends and family members that states your wishes clearly. Wear an Alcor bracelet with instructions (he shows off his own). Talk about cryonics; preach it. Put ads in the local paper. Include cryopreservation plans in your obituary.

The audience is spellbound, but he pulls them up short. "It's not easy, though," he admits. "Being outspoken like that is difficult. It makes us look weird to neighbors . . . it's strange . . . it's 'against God's will.' But the more you talk about it, the more normal it will seem."

In the end, he admits, it's a numbers game. Cryonics needs supporters to be successful. If you're the only person in your family who believes, you're a nut. But if you're surrounded by believers, you have a fighting chance. "Get more people to join," he urges us. "If you get more members, that gives you a better chance."

More also touches on a sore spot that is really the elephant in the room. In a candid moment, he admits that the science of cryonics at the moment is weak. "We need more evidence," he says, "This isn't a faith-based practice." Or, I think he means, it shouldn't be.

THE NEUROSEPARATOR

A few minutes later, during the next intermission, I'm curious to see how well they're able to get people rapidly cryopreserved, so I sidle up next to Aaron Drake, Alcor's medical response director. I'm guessing he would be the best witness to the sorts of resistance and hostility that More described. Bald, chunky, slow-talking, and reassuring, Drake reminds me of a calm, competent EMT (which he used to be in a previous life). He's exactly the sort of guy you'd want to rely on, if you're dead.

There's just one problem. Now that I'm a few feet away from him, I can't stop thinking about patient 113. John Monts. Remember him? The guy who underwent a neuroseparation. Drake was the guy who did that.

Now that Drake and I are facing each other, and now that Drake is waiting for me to say something, I'm really wishing I didn't know about John Monts. Because all I can think about is that Drake is the guy who performed Monts's . . . neuroseparation.

But Drake is polite and friendly, and he's waiting expectantly. So I ask him how much progress they've made, because when John Monts died in 2012, several days passed before his head was cryopreserved.

With vigilance, Drake says, they've been improving their response times. In the '90s, there was a 30 percent chance of getting Alcor at your bedside at the time of death. By 2012 it was 86 percent.

The best news, though, is that at some point Alcor won't need to rely on families at all. He talks excitedly about various alert technologies that are coming online. Of course there are the companies that already offer the service of sending a response if you hit a panic button. (Think of the infamous "I've fallen and I can't get up" Life Alert commercials of the '90s.) But Drake's dream is more ambitious. He sees a future of wearable devices that detect heart rate, breathing, and motion (or lack thereof) and send a signal to Alcor: "He's fallen and he really, really, can't get up."

As Drake is talking excitedly about the potential for monitoring, it seems as though he's also envisioning a new, reinvented community. Gone are the days when we all had a family and neighbors who supported us and watched over us. But now, in their place, there are devices and monitors and information. That's our new community. That's our new family.

THE FUTURE BEGINS TOMORROW

In all of this talk about science, there's one piece that's missing. It's the one piece that no one, yet, has touched on. Money.

If you're planning to be frozen for a thousand years, you need to protect your assets. You also need to make sure that someone is taking care of you, so you don't end up like James Hiram Bedford, parked in a self-storage unit surrounded by moldering furniture. And, of course, if you're lucky enough to wake up on the other end, you'd want to make sure there's enough money in your bank account to buy a cup of really strong espresso.

But how?

It just so happens that there's someone in the room who has thought about this. A lot. And he has a plan.

Rudi Hoffman steps onto the stage, sporting a ruddy complexion and green blazer. He looks and acts like he might be a used-car salesman or your local high school football coach. Perhaps in a previous life he was. But now he's the financial genius behind Alcor's success.

His brilliant strategy has been to pitch cryonics as a goal for the masses. It's not just for the megarich, he suggests archly. It's for everyone. It's for everyone in this room. It's for *you.* You just need to plan ahead.

What follows is a call-and-response litany that's worthy of a Baptist revival. "How much for a whole body cryopreservation?" he asks the audience. "Two hundred thousand dollars; that's right, folks. And how much for a neuro? Yup: ninety thousand." (A "neuro" is cryospeak for the preservation of the head only.)

Hardly anyone can afford that kind of cash, he warns. "Cryonics is an expensive business. If you think you've made adequate plans, think again." It's at this point that, channeling Nikita Khrushchev, he slips a loafer off his left (not the right) foot and pounds the podium for emphasis.

A moment later, the loafer is safely back where it belongs, and Hoffman goes on to talk about inflation, market instability, and other technical financial terms that seem to be designed to cement his credibility as a financial adviser. But then, like a seasoned salesman going in for the kill, he brings us back down to earth with the punch line. The secret, he says proudly, with the demeanor of a preacher about to reveal the way to heaven, is—wait for it—life insurance.

If you have a good life insurance policy, Hoffman says, you can use that to finance the costs of cryopreservation. You pay just a little bit every month, then the money that would normally be paid at the time of your death to your surviving family members is paid instead to Alcor. It's like taking out a mortgage on immortality. (That slogan is mine, by the way, not Hoffman's. But I think it has potential, don't you?)

Then he segues into a complex discussion of rates and plans that leaves me befuddled. The cryonauts around me are furiously taking notes, though. And the man to my left has a pocket calculator on the table in front of him, which he's prodding pensively.

"This is complicated stuff, folks," Hoffman warns us. "It's going to take a lot of money." Then: "We don't even know how much it's going to cost."

That is probably the scariest part of Hoffman's lecture, at least to judge by the expressions on the faces of my fellow audience members. There are the costs of cryopreservation, of course, which can only be expected to increase as technology becomes more refined. And the costs of storage. And the costs of keeping up with advances in technology. So these cryonauts aren't just gambling that the science will work, they're also gambling that they'll have enough money to pay for that science over the next thousand years.

CANCER GIRL VERSUS SKEPTICS

The conference is beginning to wrap up, but there's one more person I want to meet before I leave. She's a relative newcomer to the cryonics world. And yet, despite that, she's become a cause célèbre for everyone in this room.

At some point in the fall of 2012, a little red button appeared in the upper right-hand corner of Alcor's Web site. It said: HELP CRYOPRESERVE KIM SUOZZI. That's all.

This link led to a page that explained that Kim had been fighting a glioblastoma, which is a particularly nasty brain tumor that tends to victimize young adults. Her affinity for cryonics is as inexplicable as her years-long struggle with cancer is inspiring, and I want to know more about her decision to be cryopreserved. I've just heard that she is here somewhere, and I want to meet her.

It turns out I don't have to look far. Kim is sitting right behind me, at the rear of the auditorium, right next to the door. Despite all that she's been through, she somehow looks younger than the Web site photo from a couple of years earlier. She has an unruly mop of short dark-brown hair and oversize glasses that, together with cheeks that look a little swollen, perhaps from the steroids that are often used to treat brain tumors, give her an appearance of owl-like wisdom. I notice that she moves slowly and carefully, as if her body can't trust her brain, or vice versa. She speaks slowly too, with pauses and stumbles. But she's warm, funny, articulate, and someone who is impossible not to empathize with.

I hear some of her story as we chat during a break, and the rest during a short presentation. She'd been a healthy college student, she says, up until about 2010, when she began having headaches that led to her cancer diagnosis. She underwent major surgery and did well for a brief period, but then had a relapse.

She told me she got the idea of cryopreservation back in college, but then it was hypothetical. What did a twenty-year-old need with cryonics? She'd heard about the life-insurance approach to funding cryopreservation too. Still, how was that relevant to her? It turns out that it was very relevant. But after her relapse, when she decided cryonics was her best hope, she had no money and no life insurance.

Then her story appeared on the user-generated news site Reddit (where the Cancer Girl moniker appeared) and was picked up in local news outlets. Amazingly, she raised $50,000. Then Alcor got into the act and offered her a reduced price for whole-body cryopreservation ($90,000). She still needed more funding, though, and her appearance at this conference is designed at least in part to generate donations.

Given all of that interest and effort, I'm curious to know what she believes. As someone who has endured multiple treatments and tests, surely she must be skeptical of what science can achieve. Are the chances of success great enough?

When I pose this question to her, Kim's answer seems scripted: "People say there's no evidence that cryopreservation works. It's still an experiment. Well, then, if this is an experiment, I'd rather be in the treatment group than in the control group."

Fair enough. But participating in an experiment comes with burdens and hassles. This quest of hers is burning both time and money, neither of which she has much of. Is this really what she wants to be doing?

She never really answers that question, which is, I suppose, unanswerable. Instead, she focuses on the money, and she does so in a way that puts everything else in perspective.

"Is a chance of a second life worth two hundred thousand dollars?" she asks. In this audience, that's a rhetorical question, so I just nod.

"Well," she says, "I added everything up and my cancer treatment so far cost five hundred thousand. Can you believe that?"

Actually, I suspect that's a conservative estimate, but I nod again.

Then comes her punch line: "How can anyone say I shouldn't spend two hundred thousand on cryopreservation? The doctors who are telling me not to do this are the ones who told me I *should* do the usual cancer treatment. And after three years and half a million dollars, that failed. At least this might work."

Touché. Cancer Girl: 1; skeptics: 0.

Kim died on January 17, 2013. As she'd hoped, she was cryopreserved, thanks to hundreds of donations. In a thousand years or so, we'll see whether it was worth it. If you're around then, look her up.

WHEN THINGS GO WRONG

Did Kim Suozzi have the right idea? Even if cryopreservation is a gamble, maybe a gamble is better than certain death. But there are also other risks that go along with cryopreservation. And those risks have nothing to do with response times or ice formation.

The conference is ending, and all of the cryonauts are milling around a picked-over table of Danishes. Then the place at the table next to me is taken by a tiny woman who seems to disappear in an oversize purple fleece pullover and baggy trousers. Her head of silver hair pokes up from the folds like a Christmas decoration. Now she turns her bright blue eyes on me and introduces herself as Maryann. Maryann tells me she's been an Alcor member for twenty-one years, but she doesn't seem proud of that record in the way that others do when they share their history. There's no boastfulness there. She's just stating a fact.

"You know, Rudi was the one who sold me my policy, and my husband's. He died two years ago."

I express my sympathies. Then I ask the utterly bizarre question that I've learned is routine in this crowd.

Was he cryopreserved?

Maryann shakes her head and stares off into the middle distance. She shakes her head again, but still doesn't say anything. It seems like she has something to say, though, so I wait.

But as I do, I'm starting to see Maryann more clearly. She seems not just quiet, but also beaten down. Is she depressed? Sick? I've just met her a minute ago, and yet I'm already worried about her. I'm still wondering about this as she begins speaking in a low voice that's hard to hear over the white noise of all the conversations around us.

She tells me the story of her husband at a volume that's not much louder than a whisper. Her husband, Tom, had heard about Alcor and became fascinated—almost obsessed—by the idea of cryopreservation. He had been a paid-up member for years when he was hospitalized with pneumonia. His doctors said it was a minor problem, and that he'd be home in a few days. Except that he wasn't.

She wipes away a tear and stares hard at the floor for a moment. The conversation around us has died down, and people are starting to head off to their rooms. The atrium is emptying quickly, and I suggest we find a place to sit, but Maryann just starts talking again as if she hasn't heard me.

He got worse quickly, and was transferred to the ICU and put on a ventilator. She talked to his doctors, and told them about Alcor and cryopreservation. They said they'd cross that bridge when they came to it. And Maryann thought that was all she had to do.

Then things got complicated. Tom got worse quickly and had a cardiac arrest. He was resuscitated, but just barely.

Later that night the hospital called Maryann to say Tom had died. Despite years of planning and a clear understanding of Tom's wish to be cryopreserved, in that moment, she completely forgot about cryonics. She didn't think about it at all.

By the time she remembered, Tom's body had been moved to the morgue and then to the funeral home. More than four days had gone by. Then it took more than a day to arrange an Alcor visit. At that point, she realized that an attempt to cryopreserve her husband was hopeless.

Maryann's tears are flowing freely now, and I hand her a few napkins from the coffee table behind us. I put a hand on her shoulder. It's all I can think of to do. Then I tell her it must have been awful for her.

"It was. It was the worst moment in my life. I really mean that. Knowing that this was the one thing he wanted more than anything else. And knowing that it was my responsibility. And that I forgot all about it. And there's no going back. No way to do that over again."

We think about that for a while. At least, I do. I can't imagine what that must feel like. To hear Maryann tell it, cryopreservation had been Tom's dream. It had become a huge part of his life. And then, at least from Maryann's perspective, he'd missed his chance. And—again from her perspective—it had been her fault.

The atrium is almost empty now, except for us. She seems to notice this too, and checks the time on an incongruously oversize plastic sports watch. I should let her go. But there's something that's bothering me.

The Alcor membership was Tom's idea, wasn't it? She was going along with her husband? So is she still a member?

She nods uncertainly. "For now, I still am. But I think I'll let it go this year. Mostly I just came back to see some old friends. They're people Tom and I have known for years and years. But honestly, I don't want to do it. It seems like too much fuss. I'd rather just go."

Maryann says good-bye and shuffles off, leaving me thinking about what the promise of cryopreservation had brought into her life. It's difficult indeed to hear her story and think of anything positive. Hopes were dashed. And there was more guilt than one person should have to deal with. For what?

And yet I suppose that's not so different from what medicine does all the time. How many patients with cancer have I taken care of who have hoped for a cure, only to have those hopes disappear as suddenly and completely as Tom's did? And certainly I've taken care of patients whose families have felt deep, inconsolable, and seemingly unending guilt because they weren't able to support a loved one in the way that they had hoped to be able to. So Maryann and Tom's story is hardly unique to the cryonics world. And it's not fair to blame the technology that incited Tom's dream any more than it's fair to blame modern medicine for the hopes that my patients live with.

Still, I can't help thinking that this technology has introduced unnecessary misery into many lives, from the struggles that James Bedford's family faced in caring for his body to Maryann's guilt to the strife that will no doubt ensue if and when John the ordinary guy decides to shoot for immortality. At the very least, that emotional wreckage is a cost that someone needs to consider. And I have to wonder—even if no one here seems bothered—whether it's worth it.

6

Crowdsourcing Survival

"WE CAN'T DO THAT"

When eighty-seven-year-old Lorraine Bayless collapsed at the Glenwood Gardens retirement community in Bakersfield, California, on February 26, 2013, no one at the scene could have predicted the national media uproar that was about to engulf them. And they certainly couldn't have predicted that they'd become the focus of worldwide media criticism. If they had, it's very possible that they would have made different decisions.

What did they do? They didn't do anything. And that, it turns out, was what got them into trouble.

Shortly after Bayless fell to the floor in a communal dining room, a bystander used a cell phone to call 911. The dispatcher, Tracey Halvorson, instructed the bystander to reposition Bayless on the floor. Then the cell phone was passed to someone else who identified herself as a nurse named Colleen. That's when things got very strange.

We don't know everything that happened, but we do have an approximately seven-minute audio recording of that call, so we can fill in some of the gaps.

First, Halvorson determines that Bayless is unconscious. She doesn't seem to have a pulse and is not breathing. As per protocol, she advises Colleen to start CPR.

"We need to get CPR started," Halvorson says.

"Yeah, we can't do CPR," Colleen replies.

As you can imagine, there is a momentary pause as Halvorson processes this new information. You can almost hear the wheels turning as she's trying to determine whether she's heard correctly. This is, after all, someone who has called 911. Halvorson's working assumption seems to be that most people who call 911 are interested in saving a life. So Halvorson is understandably nonplussed to hear that saving a life is not currently on the menu.

It doesn't take her long, though, to realize that she has heard correctly. Halvorson is incredulous at first, but then she begins to try to persuade Colleen that, in this situation, CPR might actually be a very good idea. For instance, Halvorson points out that anyone can do CPR. Then she suggests handing the phone to someone else who is more open to considering the benefits of CPR. Then she says that Bayless is going to die without CPR.

Halvorson: Okay. I don't understand why you're not willing to help this patient.

Colleen: I am, but I'm just saying that—

Halvorson: Okay, I'll walk you through it all. We, EMS, take the liability for this, Colleen. I'm happy to help you. This is EMS protocol.

There's a break as Colleen asks someone to get a supervisor. We only hear her half of the conversation: "Can you get [unintelligible] . . . right away? I don't know where he is. But she's yelling at me and . . . I'm feeling stressed and I'm not going to do that, make that call."

Then Halvorson asks, again, whether anyone else is willing to perform CPR, and Colleen says, "We can't do that."

Halvorson's desperation is palpable now, and she runs through an increasingly frantic list of others who might be willing to perform CPR. She begins by asking, "Is there anybody that's willing to help this lady and not let her die?"

When Colleen says, "Not at this time," Halvorson asks about guests who might be nearby. Colleen declines.

Then Halvorson suggests calling a gardener. This is when it becomes clear that things are not going well. I mean, when you have to pull a guy away from his weed whacker to perform CPR, it sure sounds like you're nearing the bottom of the list, option-wise. Halvorson seems to have reached the same conclusion when, in a last attempt at persuasion, she asks if Colleen might pull someone off the street: "Can we flag someone down in the street and get them to help this lady? Can we flag a stranger down? I bet a stranger would help her. I'm pretty good at talking them into it. If you can flag a stranger down, I will help, I will tell them how to help her."

Listening to this exchange is frustrating, heartbreaking, and generally depressing. Halvorson is trying every trick in the book to get Colleen to help. She's failing, though, and she knows it.

If you listen more closely, though, you can also hear Colleen's distress. In fact, the quotes above don't do justice to the discomfort that she seems to be feeling. On the recording, it's obvious that she's confused and uncertain. It sounds as though she really believes that she can't do this. Or that she shouldn't.

It's shortly after this point in the conversation that paramedics arrive on the scene. A few seconds later, the call concludes and the recording ends. Bayless is transported to a nearby hospital, where she is pronounced dead. But the uproar is just beginning.

First, there was widespread condemnation of Colleen and Glenwood Gardens. The headlines tell most of that story. "Staff at Senior Living Home Refuses to Perform CPR on Dying Woman," a local news station proclaimed.

Expert opinion was hardly more forgiving. One physician described the lack of CPR as "horrifying." She went on to say: "I think anyone with any clinical training is morally obligated to try to help in a situation like this."

Other experts took their censure one step further. It wasn't just that a nurse has an obligation to perform CPR, they opined. Anyone does. "All of us," an ethicist is quoted as saying, "have a duty to respond to people in life-threatening situations. This is a general ethical commitment we have to each other as part of living in society."

If you read the transcript and listen to the howling of the media and their pundits, Bayless's story sounds like an open-and-shut case of bad nursing home care. In short, it sounds like the case of a nursing home that callously let its resident die. That's the simple, easy answer.

But it probably isn't the truth.

As the wave of criticism was breaking, it emerged that Colleen was hired not as a nurse but as a resident services director. So she truly was a bystander. And Bayless wasn't in a nursing home, she was in an independent living facility—essentially an apartment complex.

Now consider the fact that, by most accounts, Bayless's family was satisfied with the care that she received. Bayless, they said in a statement, wanted to die of natural causes. Although neither Colleen nor the other bystanders knew that, they nevertheless did what Bayless would have wanted, which is to say, nothing.

Then Bayless's death certificate was posted. She died not of a cardiac arrhythmia or myocardial infarction, but of a stroke. That's significant, because in a patient who collapses as a result of a stroke, CPR is unlikely to save a life. This isn't a criticism of Tracey Halvorson's advice, of course. She made all the right recommendations. But in this case, it's unlikely that CPR would have helped. And in fact local law enforcement declined to pursue an investigation of Colleen or Glenwood Gardens.

If you look at this story dispassionately, the whole episode begins to seem surreal. Lorraine Bayless was an eighty-seven-year-old woman who had led a full life and who wanted to die a natural death. Even if CPR had been attempted, it would have been highly unlikely that she would have survived. After a serious stroke, even if she did survive to leave the hospital, a substantial degree of neurologic impairment would be likely.

So why the drama about what should have been done? And how is it that we've arrived at the point at which saving a life—or trying to—is mandatory? How did it happen that all of us have "a moral obligation" and "a duty to respond"?

Not too long ago, the sorts of resuscitation techniques that could have been applied to Lorraine Bayless were still largely theoretical. Even fifty years ago, if she had collapsed, that would have been the end. But advances in the science of resuscitation have so thoroughly permeated our culture that no matter where you are, the crowd has an obligation to intervene.

And the crowd does intervene. All the time. Although it's difficult to get accurate estimates, it's possible that somewhere between 250,000 and 350,000 people suffer cardiac arrests every year in the United States. Add to that the cardiac arrests that happen in hospitals—at least 200,000 per year in the United States alone—and that's a lot of opportunities for bystanders and professionals to do CPR. (However, as we'll see, not everyone steps up to the plate when the opportunity arises.)

In a sense, the past fifty years have been a gradual revolution in democratizing resuscitation. No longer the province of medical professionals, now anyone with two arms and a sense of rhythm can resuscitate someone. It's a revolution that we're all part of, whether we know it or not.

THE MOST KISSED FACE OF ALL TIME

The notion of crowdsourcing resuscitation is an idea that's as old as resuscitation itself. The Amsterdam Society was basically a group of concerned individuals who got together spontaneously to try to keep their fellow Dutch citizens from dying. In a word: crowdsourcing.

The problem they faced, of course, was that it was difficult to generate much of a crowd if you can't offer them tools that are more effective than barrels and trotting horses. Moreover, it's difficult indeed to instill a sense of obligation and duty if what you're doing is nothing more than performance art. So if crowdsourcing resuscitation was going to catch on, someone needed to ramp up the science of CPR, bringing CPR on the street into the same league as resuscitation in hospitals.

Alas, that took a little longer than our Dutch friends might have hoped. Despite the advances of the Society, and the later discoveries of the receiving house in Hyde Park, the nineteenth century was not a particularly productive one for the science of resuscitation. In fact, the first part of the twentieth century wasn't much better. If you were a cardiac arrest victim who was "apparently dead" in 1951, your chances of survival probably weren't much better than they would have been a hundred years earlier.

Fortunately, this dismal track record finally began to improve, thanks in large part to an anesthesiologist named James Elam. Elam was a contemporary and collaborator of Peter Safar, the CPR pioneer whose Safar Center for Resuscitation Research at the University of Pittsburgh would later cause a misplaced media frenzy over "zombie dogs." In a lecture Safar gave in the last years of his life, he credited Elam with inciting his own pursuit of resuscitation: "He sparked me into a lifelong pursuit of animatology," Safar said. No pun intended.

In 1946, Elam had just arrived at the University of Minnesota Hospital in the middle of a polio outbreak that was ravaging Minneapolis. His tour of the polio ward was interrupted when a nurse and two orderlies hustled down the hall with a gurney, carrying a young boy. Elam saw that the boy wasn't breathing and that he was rapidly turning blue. (Polio can cause such profound weakness that patients are unable to swallow their saliva or keep the back of the throat open to breathe.) So Elam sealed the boy's mouth and breathed into his nose. "In four breaths," Elam reports, "he was pink."

That maneuver wasn't a new invention, and indeed many people had thought of mouth-to-nose resuscitation as well as the mouth-to-mouth variety, including crowds of concerned citizens in Europe. The problem, though, was that people hadn't gotten it to work. Simply breathing into a person's nose is unlikely to deliver air to the person's lungs. It's far more likely, instead, to go into the person's stomach, where it will do no good whatsoever, or out through the mouth, which isn't much better.

But Elam was an anesthesiologist who knew something about anatomy. And unlike his predecessors, he knew that air wouldn't reach a person's lungs through the nose unless the head is tilted at just the right angle, the jaw is pushed forward, and the mouth is sealed. So by bending the boy's neck back, and by nudging his jaw forward and sealing his mouth, Elam was able to open the boy's trachea wide enough to ventilate his lungs effectively. The boy survived, and the science of CPR was born.

Inspired by that success, Elam embarked on a series of ingenious experiments that might face some close scrutiny from an ethics review board if they were proposed today: As patients were just waking up from anesthesia—still groggy and paralyzed—they were disconnected from the ventilator. Then a physician would lean over and breathe into the endotracheal tube that went down to their lungs. Elam would monitor the oxygen in the patients' blood and was pleased to learn that exhaled air was sufficient to keep people alive. Presumably those patients were even more pleased to wake up.

Later experiments in conjunction with Peter Safar were even more ambitious. The two were convinced that mouth-to-mouth ventilation of another person's lungs could be effective, but was it practical? Could you train someone to do what they, as anesthesiologists, were able to do?

So they tried a new experiment in which twenty-five volunteers were sedated—heavily, one hopes—and paralyzed with a drug called curare, a natural version of many of the medications used today in surgery. A total of 167 people were then brought into the room and watched either Elam or Safar perform mouth-to-mouth ventilation on the volunteer. Finally, when the laypeople seemed to have the hang of the procedure, they were given a chance to try it for themselves.

All concerns about personal hygiene aside, what Elam and Safar did was remarkable. In a sense, they took the real-world experimentation of the receiving house in Hyde Park to the next logical step. Not content simply to observe what happened, they brought the resuscitation process back to a laboratory where it could be monitored, measured, and controlled. And their experiments would prove to be hugely influential in shaping the course that resuscitation science was to take.

Elam and Safar deserve a lot of the credit for getting resuscitation to where it is today. But not all of it. Because like everyone who had come before them—or almost everyone—they were still missing a key piece of the resuscitation puzzle. They needed to get the heart pumping too.

That happened in 1958 when a group of researchers stumbled on this problem. Three researchers at Johns Hopkins—William Bennett Kouwenhoven, Guy Knickerbocker, and James Jude—were studying the effects of external shocks on dogs' hearts. Somewhere in the middle of an experiment that was probably about as gruesome as it sounds, they realized that the act of placing the defibrillator paddles on a dog's chest created a faint but noticeable pulse.

Imagine their surprise. That sense of surprise was probably exceeded only by that of the dog when he discovered that the rules of the game had changed quite dramatically and that now three well-respected scientists had begun to pound on his chest. Even in the already-difficult life of a canine research subject, that probably would have qualified as a very bad day.

Within a year, the trio was trying out their new technique on people. And one of the first was a thirty-five-year old woman who suffered a cardiac arrest just before undergoing gallbladder surgery. She lived.

So there you have it: the building blocks of CPR. More than fifty years ago, we'd figured out how to breathe for a dead person, and how to pound on his chest. In short, we knew how to keep someone alive, at least for a little while.

But while it's one thing to revive a dog in a lab, the real challenge was resuscitating people in the real world, far outside the familiar confines of an operating room. Safar and Elam and others realized that they needed to teach the skills that they were perfecting to bystanders so that if anyone—anywhere—attempted to become late, there would be someone right there who could help. It's one thing to train a handful of volunteers in an operating room but another thing entirely to train a CPR army.

Although no one realized it at the time, the solution was born back in the late 1880s, when the body of a young woman was pulled from the Seine in Paris, at the Quai du Louvre. There was no sign of violence, so suicide was the suspected cause of death. That, by itself, was hardly noteworthy. By some estimates more than a hundred women ended their lives this same way every year. That river running through the heart of Paris was, unfortunately, an all-too-obvious means of suicide.

What was noteworthy, though, was that a pathologist at the Paris morgue was so enchanted by her face—even in death, and after what was probably days in the water—that he had a molder make a plaster death mask for him. That mask, or others like it, became wildly popular in the Bohemian salons of the late nineteenth century. The mask inspired comments from Albert Camus, for instance, and a poem by Vladimir Nabokov ("L'Inconnue de la Seine," originally written in Russian), among others. Copies of the original were common, and imitations were numerous.

Almost a century later, in Stavanger, Norway, one mask fell into the hands of Asmund Laerdal, then the owner of a toy manufacturing company. Coincidentally, Peter Safar had given a lecture about his research in September 1958 at the meeting of the Scandinavian Society of Anesthesiology and Intensive Care Medicine. A Norwegian anesthesiologist named Bjørn Lind was in the audience, and, inspired by the possibilities for public education, approached Laerdal about making a model on which members of the lay public could be taught how to perform resuscitation. After contemplating the mask he'd discovered, Laerdal used that anonymous woman's face to begin manufacturing a life-size mannequin who has since become ubiquitous in medical schools and classrooms.

You probably know her as Rescue Annie. Or Resusci Anne. And if you've met Annie, then you've kissed the same face that was pulled from the Seine more than a hundred years ago.

Annie and I first met years ago, in a medical school classroom.

But it's been a while, and I'm thinking this might be a good time to get reacquainted.

PUSH HARDER. PUSH FASTER. REPEAT.

Just as she was the last time we met, again Annie is lying on the floor. As usual, she is dead.

Admittedly, it's difficult to tell whether a mannequin like Annie is dead. For instance, mannequins don't scratch their chins or sneeze or wink at you. They don't do much of anything that can reliably distinguish the quick from the dead. This makes it difficult to determine with certainty that they are, in fact, dead.

This ambiguity, in turn, makes mannequins not particularly well suited to a CPR training class, which is where I am right now. I'm in the basement multipurpose room of a suburban middle school, standing in an uneasy circle of a dozen fidgety adolescents with the attention spans of fruit flies. None of them seems particularly concerned about Annie's future health and well-being.

Nevertheless, we know that she's dead because there is a large, slightly intimidating man standing over her, who has announced this fact. His artfully mussed sandy hair, crisp chinos, and polo shirt make him look like a gym teacher, which is exactly what he is. The cause of death, Mr. Gym Teacher tells us with a macabre enthusiasm, is a heart attack.

"What do you do now?" Mr. Gym Teacher asks the group.

Mr. Gym Teacher turns to the person next to me, a girl of maybe thirteen. He looks at her intently and enfolds his barrel chest in bulging forearms, which, he confided to me when I arrived a few minutes early, came from his evening work as the drummer for a Journey tribute band. Now he attempts a comforting smile that, in this context, comes off as just plain scary.

"What. Do. You. Do?"

Silence. Deer. Headlights. There is an intensely awkward moment as the girl looks very intently at her Converse-clad feet.

But fortunately for us all, in that moment, a future doctor is born. On the other side of our little circle, a hand rises tentatively. The hand is attached to a boy who is looking excited and nervous.

"You check for danger," the boy says.

Mr. Gym Teacher nods curtly. "Right. Then what?"

The girl next to me decides that her feet are perhaps not as fascinating as she'd first thought. She looks up at me. Then at Mr. Gym Teacher.

"Then you check for a response," she says quietly.

"Right," Mr. Gym Teacher says. He smiles. Everyone relaxes, just a little.

"Then?"

No one's talking, so I help out. Just a little. You send for help, I suggest, and Mr. Gym Teacher nods, pleased to be done with the preamble.

Mr. Gym Teacher is about my age, which means it's likely that he's been teaching this stuff for a while. And, like me, he's probably old enough to remember the good old days when the mnemonic we used was A (airway), B (breathing), C (circulation).

But not anymore. Now it's D (check for danger), R (check for a response), and S (send for help). In fairness, these letters seem to be proving themselves to be excellent guides. They've gotten this group of middle school students (plus one doctor) through D, R, and S. Not bad.

Next, there are our old friends A, B, and C, albeit rearranged as "CAB," which gives primacy to restoring circulation (C) through chest compressions. That, science is beginning to understand, is the most important part of the whole resuscitation thing. This is good news for many heart attack victims, but bad news for many trainees who are now stuck having to memorize the rather disgusting mnemonic "DRSCAB."

So now that we're done with DRS comes the hard part: C.

It's safe to say that we've all been dreading C. That's because we inherently dread another letter that is not part of the mnemonic—V—which stands for "volunteers." As in: "I need a couple of volunteers to demonstrate CPR, in front of their peers, while being critiqued in front of said peers by an intimidating gym teacher."

But Mr. Gym Teacher doesn't ask for volunteers, which is good. Instead, he points at the two kids who have volunteered answers so far, which isn't. Perhaps he's operating under the assumption that no good deed should go unpunished. He points at them, and then at Annie. Then, just in case it wasn't clear, he points at each of them again.

They both look very hard at the floor. The boy fidgets. The girl is starting to hyperventilate so forcefully I'm concerned that she, too, may join Annie on the floor.

Thinking back to my own adolescence, it's difficult to imagine a scenario that would be more embarrassing than performing CPR for the first time in front of a sniggering group of your peers. Unless, perhaps, it's doing so with a person of the opposite sex.

The boy—not so shrewd—takes a step forward. The girl, demonstrating a level of maturity far beyond her years, takes two steps back. She looks as though she's seriously contemplating feigning a seizure. Out of sympathy, I raise my hand.

Mr. Gym Teacher—who, for the record, amusingly enough, is named Jim and is actually a really nice guy—smiles at me and nods. So the boy and I approach Annie.

Fortunately, I've read the official manual on "How to Pretend to Revive a Fake Dead Plastic Person." (In case you, too, want some fun bedtime reading, this volume's actual title is "Part 5: American Heart Association Guidelines for Cardiopulmonary Resuscitation and Emergency Cardiovascular Care.") For what it's worth, I prefer my version.

First, I remember, the rescuer should begin by "tapping the victim on the shoulder and shouting at the victim." Tapping is easy. I tap. Annie doesn't respond.

The shout is more difficult. I settle for an innocuous but simple "Hey!"

As expected, this outburst has no discernible effect on the armless torso in front of me. Oddly, though, the boy across from me is staring at Annie intently as if maybe he was thinking she might wake up. Alas, she does not.

Next we need to start CPR. But the kid is now one step ahead of me. He remembers what, in the heat of the moment, I've forgotten. He turns, with a stage actor's gravity, to one of the kids in the circle around us and tells him to call 911. The solemnity of the moment is marred somewhat when his voice cracks into a squeak, but no one seems to notice. Certainly not Annie.

Right. So now someone is sending for help. In the meantime, it's on to CPR. This is when things get less fun, and I'm already wishing I hadn't volunteered for this duty.

The 2010 American Heart Association guidelines say that untrained bystanders can't reliably determine whether a patient has a pulse. Checking for a pulse also takes time, which the victim generally doesn't have much of. Thus the guidelines recommend avoiding this step entirely.

If someone collapses and is not breathing (or is only gasping), the guidelines say, it's safe to assume that they've suffered a cardiac arrest. Even trained health-care providers shouldn't spend more than ten seconds checking for a pulse. Lay rescuers aren't supposed to check for breathing, either, for the same reasons. It's the compressions, and specifically the rate of compressions, that matter. The more compressions you do, the more likely a patient is to recover a heartbeat and a pulse.

Unsure whether I'm a layperson or a trained health-care provider in this scenario, I skip the pulse check and the breathing check.

As I kneel down next to Annie, my back creaks a subtle but unmistakable warning. It is telling me in no uncertain terms that regardless of whether I'm an untrained bystander or a trained health-care provider, I'm most certainly not a twelve-year-old boy.

Nor am I as clever as this particular twelve-year old boy, who has slyly taken a position at Annie's head, leaving the rest of her to me.

What that means, in practical terms—and these terms are the only terms that my aging back cares about right now—is that while I'm doing thirty chest compressions, this kid is just going to sit there. Because that's the ratio we're expected to provide. Thirty compressions for every breath.

What's even more unfair is the fact that he's not even essential to this CPR effort. There's growing evidence that maybe chest compressions are the *only* part of CPR that matters. In one meta-analysis of adults with out-of-hospital cardiac arrests, compression-only CPR by the lay public had a higher success rate than standard CPR did. That is, cardiac arrest victims may actually do better if someone isn't crouched over them trying to breathe for them. The reasons for this aren't entirely clear, but it may be that it's just too much to ask untrained bystanders to do chest compressions and breathing. It's a little like talking on a cell phone while you're driving. If rescue breathing distracts rescuers from doing (much more important) chest compressions, then the outcomes would be worse. Maybe much worse.

Anyway, the guidelines—perhaps written by the same Madison Avenue firm that penned Nike's "Just Do It" tagline—admonish rescuers to "Push hard and push fast." So I do.

Alas, apparently I'm not pushing fast enough. Jim asks me, in between compressions, how fast I'm supposed to be . . . compressing.

It's amazingly difficult to answer a question like that when you're bouncing up and down like a demented bobble toy. Especially when your back is giving you the warning signs of a crescendo of twinges that indicate it's not going to stand for this abuse much longer.

The answer, I know, is one hundred compressions per minute. And each compression should depress the chest wall by two inches. And that's what I'm doing. Or, at least, that's what I think I'm doing.

"Your depth is good . . ." Jim reports. Wow. My depth is good. I'm a star. I grin.

And seriously, that's something to be proud of. One large study of out-of-hospital cardiac arrests found that rescuers don't usually get the compression depth right. That's important because the same study also found that patients who didn't get the right compression depth had worse chances of survival.

"But you're not going fast enough." There's a moment of silence in which I try to think of a clever retort, without success.

"Think about the beat of 'Stayin' Alive,'" Jim says. "You know, the Bee Gees."

In the few seconds it takes me to look up at Jim and nod, I register a wall of blank looks around me. I swear I can see wheels turning slowly in little adolescent heads. I also see a few of them mouth the words as if they're trying out words in a foreign language.

What are Bee . . . Gees? they ask themselves. What indeed?

It's true, that iconic song from 1977 provides the perfect beat for chest compressions. Unfortunately, this is probably not very useful information. In the year 2014, the only person at a typical cardiac arrest scene who is likely to appreciate this rule of thumb is the elderly person who is lying on the floor.

Life is not fair to the aged. Not only am I being forced to do thirty laborious chest compressions for each little breath that the kid delivers, but I'm also doing it at the exhausting rate of a hundred beats per minute. I'm thinking this would be an excellent time to remind everyone around me that rescuer fatigue sets in very fast—after the first minute of CPR. Studies show fatigue results in a measurable decline in CPR effectiveness after the first minute of compressions, even though rescuers may not feel tired.

For the record, though, I feel tired.

Meanwhile, the kid is clowning and making a show of his exhausting one breath given every thirty compressions. The crowd begins to giggle. I keep sweating.

I give Jim a warning look. He relents. "Switch," he says.

I smile and try to straighten my back, which I think is permanently locked into a C-curve. The kid looks uneasy. I smile sweetly and body-check him out of my way, pointing at Annie's torso.

Now I have to do B for breathing. At a rate of about three breaths per minute. Each breath gets about one second. And each breath should have a tidal volume of 500 to 600 cc. This is about the volume of air that's in an empty Venti iced latte cup, once you drink its contents. Something I'd really like to do right now.

Instead, I tilt Annie's head back, pinch her nose with my left hand, and move her jaw forward with my right. I've done this perhaps half a dozen times with a real person, but with Annie, there's much more resistance and it's more manhandling than there would be under normal circumstances. Annie doesn't seem to mind.

One breath, and Annie's chest rises and falls. Easy. Then I sit back and watch happily as the kid proceeds to get everything wrong.

His compressions are not deep enough, which one of his classmates is quick to point out. Then he gets flustered, and his compressions slow down. He is not anywhere close to "Stayin' Alive" speed.

I mention this.

The kid is momentarily distracted by my helpful advice and he pauses, which everyone knows you're never, ever supposed to do.

"Push harder!" someone yells.

"Push faster!" another chimes in.

Soon they have a singsong pep rally chant going.

"Push harder!"

"Push faster!"

"Push harder!"

"Push faster!"

Converse Girl and the girl next to her are slapping their thighs to keep the rhythm going. This, I'm thinking, is much better than disco. Take that, 1970s.

Alas, Jim assumes an expression that suggests he is about to put an end to this rally before it turns into a rave and Annie gets trampled. The chanting subsides gradually. Then our little rescue comes to a close.

Jim announces that we have—against all odds—saved Annie. She's alive. Except for the fact that she's still missing two arms, two legs, and pretty much everything else.

Nevertheless, the kid and I are heroes. Jim invites everyone to clap for us, and, much to my surprise, they do. What's more, I feel good. I really do. I saved Annie!

The kid, back in clown mode, takes a bow. I contemplate joining him, but my back suggests that any sudden movements would not be welcome right now. Instead I hobble out of the circle to safety.

Finally, after it's all over, there's a reward. Each of us gets an I LEARNED HOW TO SAVE A LIFE TODAY! button. I'm thinking this might be worth a free drink at my local bar. And since I'm the only one here who is old enough to take advantage of this potential perk, I seriously consider asking a couple of these kids to hand theirs over. Converse Girl, at least, owes me.

"MAGIC FINGERS" CPR

Several weeks later, my back is starting to feel more normal. But the memory of my encounter with Annie is still fresh in my mind, and I still get a reminder of that afternoon whenever I lean over to put a leash on my dog. CPR may be effective, but it's not easy.

But at least the theory is simple, right? You push on the chest of a dog—or a person—and that pressure squeezes blood through the heart. What could be more straightforward?

Actually, the mechanics of CPR aren't nearly as simple as they sound. Josh Lampe, the engineer who was testing various ways to cool Petunia the pig while her heart was stopped, explains to me that although CPR looks like something even a chimpanzee—or an eighth-grader—could do, it turns out that it's actually far more complicated than you'd think.

We're talking about the *C* part of CPR, and Josh is holding up a piece of paper in one hand. This, he's telling me, is the argument for the complexity of C. Sprawled across the page is a graph that looks to me like a stream of squiggly lines layered on top of one another.

"This," Josh says dramatically, "is a pig."

What he means is that these lines are various measurements made in a pig who was undergoing CPR. These squiggly lines, he says, represent a composite picture of the way that blood flows through a pig—and presumably a person—while someone is pounding on its chest.

These lines tell us how CPR works, and how those chest compressions move blood through the pig's heart and out into its circulation.

To me, these squiggles look like a generous slice of baklava in cross-section. To Josh, though, each line tells a story, and he can read the whole mess in much the same way that a surveyor reads a topographic map. What he sees is both confusing and really, really cool. And if you're an engineer who makes a living studying the insides of pigs, those two adjectives amount to pretty much the same thing.

The story told by these squiggles flies in the face of what I, at least, was taught in medical school. I'd always figured that when I performed CPR on a patient, the mechanical force of the chest compressions re-created some semblance of the heart's normal function. Because the heart's valves allow blood to move forward but not backward, a little squeeze, I thought, would produce something like natural circulation. Whenever I've been called to a cardiac arrest and ended up responsible for C, I've always thought of the mechanics of one of those small foot-operated pumps you use to inflate an air mattress. That is, I've thought of myself as a large foot.

However, the squiggles in front of me suggest that the explanation is not that simple. As Josh points out, the blood flow illustrated on this graph is minimal. Blood isn't really circulating at all.

That's eye-opening. It's also a little depressing. Is he saying that all of those chest compressions I did as a medical student and resident didn't really push blood out of the heart and into the circulation?

Yes, it turns out that's exactly what he's saying. In this pig model, at least, the blood is just washing back and forth. This is known in technical circles as the sloshing theory. Maybe, the theory goes, chest compressions don't really move blood from veins, through the lungs, and out through the aorta. Maybe they simply push blood out in both directions—into the veins and into the aorta—and then suck it back in. This possibility is particularly interesting because it suggests that if CPR works, it doesn't necessarily require the difficult task of making blood flow normally. Maybe it's enough to simply slosh blood around.

To see how that might work, remember that the chief goal of

compression is to get oxygen to key organs like the brain. One way to do that is to load oxygen into the red blood cells as they pass through the lungs, and then push them out to waiting organs. But sloshing could do much the same thing. When you slosh blood around, you mix it. And when you mix any liquid, the stuff that it's carrying diffuses, and concentration gradients eventually disappear.

Put a few drops of blue food coloring into a bucket of water, and it forms a little blue cloud. But if you jostle the bucket repeatedly, that cloud starts to spread. Pretty soon you have a bucket full of blue water. That's the appeal of the sloshing theory. Get a little oxygen into the blood via the lungs, and then slosh it up to the brain and everywhere else it's needed.

This seems farfetched, but Josh points out that there are other situations in which sloshing seems to work quite well. For instance, there is an odd type of ventilator that is occasionally used in ICUs called a jet ventilator. It's used when the lungs are very stiff or severely damaged, and when normal (large) in-out breaths could damage lung tissue. Instead, the jet ventilator provides very small, very frequent puffs of air that diffuse oxygen at low pressure. Watching one of these in action, it's difficult to believe it could deliver any oxygen at all, but it does.

So is it possible that more effective sloshing could lead to more effective CPR? Maybe not. Blood is much more dense, and so you can't slosh blood as quickly and effectively as you can slosh air. And with CPR, you're talking about an entire body full of blood, not just a small bag of air. Nevertheless, there is an odd line of research that's capitalizing on the sloshing theory.

If you've ever spent any time traveling the back roads of the Midwest, and if you've ever stayed in an old motel—the kind that proudly advertises "Color TV" and "Air Conditioning" as if they're newfangled inventions—you may have encountered an odd device that was popularized in the 1970s called a Magic Fingers bed. Drop a quarter into the metal control unit on the nightstand, and the mattress begins to vibrate, providing fifteen minutes of "tingling relaxation and ease." Or so the sign promises. Having tried this device once in a motel outside the tiny town of Newberry in the Upper Peninsula of Michigan, all I can say is that it feels vaguely disorienting in a numbing sort of way. Like enduring a particularly long, mild earthquake.

Perhaps inspired by an evening spent at just such a motel, two physicians, Jose Antonio Adams and Paul Kurlansky, have developed a table that does the same thing. That is, it oscillates back and forth and, they claim, it moves blood in the same way that CPR does. It may also move air through the lungs.

This technique, alas, has not really caught on. In part, this lack of enthusiasm is due to a lack of evidence that it works. In addition, during a cardiac arrest, just plopping a patient down on a wiggling bed or table doesn't seem like much of an intervention. For doctors and nurses, though, an oscillating bed is no doubt a welcome sight. After an hour of laborious CPR, I can imagine code teams drawing straws to see who gets a Magic Fingers massage first.

"LIKE, JUST PUSH ON HIS CHEST"

So the mechanics of CPR are more complex than you might think. But while it might be true that one can push on the chest to get oxygen to the brain, it's another thing entirely to get a person to do that pushing. Obviously, even the most sophisticated studies of how CPR works won't do anyone any good if you can't induce bystanders to "push hard and push fast." Or to push at all.

Remember the story of Lorraine Bayless and Colleen? Colleen's reluctance to intervene is hardly unique. Every day, witnesses to a cardiac arrest watch a person collapse, notice that the victim is no longer breathing, and . . . do nothing. As you'd expect, very few of the stories that illustrate this inertia have happy endings, but there is one that did.

When Tristin Saghin discovered his sister Brooke facedown in the family pool, he pulled her out of the water, laid her out on the concrete, and started CPR. Chest compressions, breathing . . . he did exactly what he was supposed to. His sister began to breathe again on her own, and soon she woke up. Eventually, she made a full recovery.

Nothing surprising there, right? That's the way that CPR is supposed to work. And that's what we're all supposed to do when confronted with someone who is at risk of becoming dead.

The odd thing about this story, though, is that Tristin was only nine years old. His sister was two. Tristin had never taken a CPR course. Neither did he have any idea what he was doing.

He had, however, watched the film *Black Hawk Down*. And in one scene, as Tristin recalls it, "they were, like, pushing on [his] chest and giving him rescue breaths." So that's what Tristin did. He pushed on Brooke's chest and gave her rescue breaths, and she survived. That success led to a media blitz and, like, a congratulatory e-mail from *Black Hawk Down*'s producer, Jerry Bruckheimer.

So life imitates art, and lives are saved. All is well, right?

Not entirely.

It's true that Tristin saved the day, but it's strange that he had to. For instance, Tristin's mother and grandmother were right there. They could have started CPR, but they didn't. So it fell to Brooke's nine-year-old brother.

How could two adult bystanders stand there, doing nothing? Even if the mechanics of how CPR works are complex, the key steps in performing CPR are really quite simple. "Push hard and push fast" pretty much sums it up. And yet these two intelligent, caring adults didn't do either.

This is hardly unique. For instance, one study found that bystander-initiated CPR occurred in fewer than one-third of cardiac arrests. That doesn't mean that one in three bystanders stepped up to the plate. That wouldn't be so bad. If you're lying on your back next to a pool with no discernible signs of life, you don't need an army. You just need one person, with two arms and a sense of rhythm.

No, what it means is that two-thirds of the time, no one stepped up to the plate. No one, sometimes, out of a crowd of ten or more. That's a lot of spectators in the stands, and not nearly enough people out on the field where they can do some good. What was even more troubling was that this study found bystanders' enthusiasm varied by neighborhood. Bystanders in lower-income neighborhoods, for example, were even less likely to help.

But why?

There are many reasons why bystanders like Colleen and Brooke's mother and grandmother don't help. People are uncertain about what a cardiac arrest is, for instance. They don't know the procedure for CPR, and they're afraid of doing CPR incorrectly and either hurting the victim or facing liability, or both. There is also the ick factor—and a deep-seated reluctance to use your mouth on someone else's, particularly when that person is a perfect stranger. We also are likely to wait for others to act first.

All of those reasons help to explain why we tend to be reluctant rescuers. On the other hand, you have to wonder why we let ourselves be dissuaded by these reasons. Because if you look around, we're surrounded by examples of successful CPR.

Indeed, it seems as though every day the headlines trumpet stories of heroic CPR efforts and lives saved. We hear about the man who has a cardiac arrest on a golf course and is resuscitated by a caddy. Or the woman who collapses while power-walking in her local mall and is revived by a clerk at Sears. These stories are everywhere, giving the impression that any one of us could be a hero.

And we get the same message from films and television. If Tristin Saghin's story illustrates how life can imitate art, then art exaggerates life. I'm thinking in particular of the way that television portrays CPR. In a classic study, a group of researchers watched episodes from three television shows that were popular in the 1990s and recorded instances of cardiac arrest and CPR. What they found was that 75 percent of patients who underwent CPR were revived, and that about two-thirds appeared to survive long enough to leave the hospital.

Now, at this point it should be made clear that real-world cardiac arrests don't always end so well. Those numbers are two or three times better, at least, than what nontelevised patients could expect in the 1990s. And many of the television "patients" experienced their cardiac arrest outside the controlled setting of a hospital and would have done much, much worse.

So we're exposed to these stories of spectacular successes all the time. We see reports of stunning rescues portrayed in headlines, on television, and in movie theaters. The message—clear and unmistakable—is that CPR saves lives. We should all learn CPR. And we should always, always perform CPR when we have a chance, because any one of us could save a life. And yet most of us don't.

But hope isn't lost. Maybe we'll never truly crowdsource CPR, but that's OK. Maybe we don't need to.

MS. HOT SAVES A LIFE

There is a man lying on the floor. There is a certain family resemblance to Rescue Annie and, like her, this man is very clearly made of plastic and metal. So no one in the room is particularly distressed by his deadness. There is one person, though, who is doing a passable imitation of someone expressing concern.

Leaning over the dead mannequin is a young woman of perhaps twenty-five. In a gesture that is borrowed from either television medical shows or the adult film industry, she rips the mannequin's shirt open, exposing his plastic chest. And then, in a gesture that is not taken from any medical television show that I've ever seen, she brushes long blond hair back with both hands, and sticks out a chest that is emblazoned in silver glitter with the word "hot."

Ms. Hot and I are in a newish, linoleum-tiled windowless room buried deep within the bowels of a suburban shopping center that has recently embarked on an effort to train its employees in the use of eight new automatic external defibrillators (AEDs). These are devices that let bystanders with no medical training shock a heart back to life. The harsh overhead fluorescent lights illuminate a group of ten people, all of whom are shopping center employees. And we're all here for a crash AED course.

Now Ms. Hot has dispensed with her theatrical *Vogue* pose and has turned her attention to the AED that is next to her. It's basically a stripped-down version of Adam, the self-important defibrillator we met in chapter 3. But while Adam was designed for use by medical professionals who could choose whether to follow his advice, this version—let's call her Eve—is not going to permit any improvisation. She's hardwired for safety and can't be overridden, modified, or cajoled into giving a shock when she shouldn't.

Ms. Hot removes two patches from Eve's front pocket and places them on the mannequin's recently exposed chest, at roughly nine o'clock and three o'clock. Ideally, the patches should be a little more north-south, and Ms. Hot's placement is optimally suited if her goal is to defibrillate his spleen. Oh, well.

Ms. Hot presses a button on Eve's face. Then she waits as Eve reads the patient's heart rhythm.

"The rhythm is ventricular tachycardia," Eve says in a tone that indicates she's seen it all before.

"Stand clear," she warns.

Ms. Hot, perhaps fearing that the room will be engulfed by an electrical storm, jumps to her feet, and takes half a dozen steps back. Suddenly I'm actually closer to Eve and the patient than anyone else, and all of the women in the room are looking at me worriedly. I'm pretty sure there's no immediate danger of electrocution from six feet away, but I take a step back too. They look relieved, and turn their attention back to the mannequin.

Eve winds up and delivers a shock. Ms. Hot seems somewhat confused. It's as if she were expecting sparks and singed flesh.

"The rhythm is sinus," Eve reports. Then she checks out.

The instructor thanks Ms. Hot for a job well done, and everyone applauds. Ms. Hot takes a bow. Just another day of life and death in suburbia.

I've just met the most exciting development in crowdsourcing CPR to come around in the past twenty years and its star is . . . you. Or me. It's all of us.

Just as CPR revolutionized the science of resuscitation, taking an OR procedure to the streets, AEDs take the lessons of defibrillation to malls, stadiums, and even airplanes. Now any one of us can slap a couple of patches on a chest, yell "Clear!" and perhaps even save a life.

To be fair, at least some of the technology is relatively simple. Remember Charles Kite? The DIY paramedic who is credited with shocking a girl back to life after a fall in 1788? He accomplished that feat—the story goes—with a homemade battery. Basically he used a charge of electricity in a jar. So the science of shocking someone is not terribly complex or new.

What is complex and what delayed this latest advance in crowdsourcing by more than two hundred years, is figuring out when to shock. Some rhythms respond to a shock, and others don't. As we've seen, shocking asystole (a flat line) won't work, no matter what fictional rescues you may have observed on the television screen or in movie theaters.

The real risk of a mistake, though, arises if you shock a normal or almost-normal rhythm. A sinus tachycardia, for instance, is a normal rhythm, propagated in the usual way, but is just faster than normal. Apply a shock to that rhythm, and you could cause ventricular fibrillation, which would be bad. Or asystole, which would be worse. (Ms. Hot's anxiety notwithstanding, the risk to a bystander is negligible.)

Now, maybe I'm being unfair, but if that were me, flat on my back without a pulse, I'm not sure I could count on someone like Ms. Hot for an accurate diagnosis. So any DIY resuscitation device needs some way to figure out whether a rhythm will respond to a shock. And it needs to be able to restrain itself from delivering a shock that would be ineffective, or even harmful.

The idea behind AEDs has been around for a while, mostly as a science-fiction dream, reminiscent of those *Star Trek* gadgets that would diagnose injuries, treat them, and give you a deep tissue massage, all at the same time. But the progression of their development has been gradual. Early defibrillators provided shocks only, and later became integrated defibrillator-monitors. At first, those integrated AEDs offered cues and suggestions to health-care providers, and from there it was a relatively short step to offering more simple recommendations to bystanders.

That's what Eve did for Ms. Hot. Eve was a semiautomated device. Eve read the rhythm and then asked Ms. Hot to push a button to administer the shock. I suppose it's the machine world's way of making us humans feel as if we're still in control. It's a nice gesture, but in general, in my experience, when an AED as authoritative as Eve tells you to push the button, people generally do exactly that.

These AEDs are undeniably cool. They can make a paramedic out of almost anyone, including middle school students. But do they work?

Indeed they do. In one large descriptive study of out-of-hospital arrests, patients who received bystander CPR had a 9 percent chance of survival, whereas those who had an AED applied had a 24 percent chance of survival. What was most impressive, though, was that the AED group's survival increased to 38 percent if the AED found a rhythm that it could shock. And 38 percent for an out-of-hospital arrest is about as good as these numbers get. So yes, even in the hands of someone like Ms. Hot, AEDs save lives.

So should we deploy AEDs everywhere? Well, maybe not. Remember that AEDs work best if they can find an abnormal rhythm to shock. And it turns out that, for some reason, these "shockable" rhythms are more common in public places. When a bystander applied an AED in a home, 36 percent of patients had a shockable rhythm, compared to 79 percent when a bystander applied an AED in a public setting.

Why that's the case isn't clear, but it's possible that cardiac arrests at home are caused by different events (such as a massive blood clot in a pulmonary blood vessel) that are more likely to go straight to asystole, bypassing treatable rhythms. Not surprisingly, in the same study, survival rates in public settings were almost three times greater (34 percent versus 12 percent). The bottom line, therefore, is that although it probably makes sense to put an AED in a mall, it's probably not worth investing in one for your living room.

So in order for AEDs to save lives, they need to be common in public places. Common enough so that there's always one within shouting distance. Think about how ubiquitous ATMs are, and that would give you an idea of how many AEDs we need out there. The problem, though, is that we have a long way to go.

"LOCATION, LOCATION, LOCATION"

In matters of life and death, as in real estate, location matters. To see just how much it matters, I'm taking a walk. Two walks, actually, about ten miles apart.

First, I check out an outdoor shopping center in a nice suburban neighborhood. I pass a long list of nice stores selling clothes, shoes, and very expensive espresso machines.

After perhaps fifty yards, I see it. There, next to the restrooms, is an AED station. I keep walking. I turn and go into a large department store and wander through to the back. And there, right next to the customer services desk, is another AED station. Out the back door and past the Apple store. Then through a long arcade and out the other side, next to a small food court. And there, just where I was expecting it, is—you guessed it—another AED station. I'll spare you the details, but I spent about two hours in that place and came across no fewer than six AEDs.

My second walk takes place in a different sort of neighborhood entirely. Now I'm walking down a slightly iffy street. It's bordered by the University of Pennsylvania on one side, and on the other side by a neighborhood that gets dicey very quickly. To my left, each subsequent block contributes a progressively higher prevalence of boarded-up buildings, overgrown lawns, and other indicators of decline.

I'm keeping to what feels like the safer side of that dividing line, and again I'm on the hunt for AEDs. I wander by a gas station, and then another gas station. Nothing.

Then my hopes rise as I spy a convenience store. I wander in, picking up a package of M&M's for company. Nope. No AED.

A block farther on, there's a small strip mall of perhaps a dozen stores. Maybe this is the place? A hairdresser, a liquor store, a Radio

Shack . . . I peer into windows and stop into a few stores just to be sure, but it's as I suspected: no AEDs.

The problem with AEDs is that they're not everywhere. My training session with Ms. Hot is living proof of that. That suburban mall had made a commitment to ensure that no one was going to die of a cardiac arrest on its watch. That's admirable. Ditto the second shopping center I visited. And other establishments have made the same commitment. Hospitals, certainly. And airports. Even airplanes.

What's interesting is not where these AEDs pop up, but where they don't. Dr. Raina Merchant, a colleague of mine at Penn, has been studying the way that AEDs map to populations, and it was her research that prompted my two strolls today. In Philadelphia, for instance, she's found that there are more AEDs per capita in areas of the city where the population is predominantly white, well-off, and highly educated. Conversely, if you're in a poorer, largely African American neighborhood, you probably shouldn't count on an AED coming to your rescue. If you're thinking about having a cardiac arrest, a suburban mall is an excellent place to be. The corner of Forty-Ninth and Market Street in West Philadelphia, not so much.

Don't like that uncertainty? Well, you know what they say: if you want something done right, you have to do it yourself. That's perhaps more difficult when it comes to restarting a heart than it is, say, if you're making a cup of coffee or changing a tire. But the same principle applies.

But how can someone restart their own heart? I'm about to meet someone who has pulled off this trick not just once, but several times. Although, truth be told, he's had a little help.

SHOCK IN A BOX

John died yesterday. At least, he looked pretty dead to his buddies on the seventh green of the suburban golf course where he'd been playing just a few minutes before. In the space of a couple of seconds, his handicap went from a respectable 14 to something close to infinity.

But then he was back.

And now he feels fine.

John didn't endure the indignities of his golfing buddies pounding on his chest or providing mouth-to-mouth resuscitation. Even better, he didn't suffer any of the usual CPR side effects of a bruised chest or cracked ribs.

That's because his buddies didn't do CPR. In fact, they didn't do anything. For all I know, one of them snuck in a surreptitious short putt, hoping no one would notice the extra stroke.

The truth is that John's buddies did exactly what they needed to do. Nothing.

John is a big, affable guy in his late fifties. He's easily six feet tall, with close-cropped gray hair, broad shoulders, and a dense web of freckles that he's earned through years of outdoor work as a commercial construction contractor. I'm meeting him in a clinic that specializes in the care of people who have a tendency to die suddenly. That is, they all have a history of ventricular fibrillation and other fatal arrhythmias. With expansive gestures, he's telling me about what happened yesterday. But his story, he says, actually began much earlier.

Two years ago, he'd been sitting in the bleachers at his grandson's first Little League game. It had been hot that afternoon. He started feeling woozy but chalked it up to the insufferably muggy Philadelphia summer weather.

"Then the next thing I remember," he says, "I'm in an intensive care unit. I'm lying flat on my back, with these straps tying my arms and legs down. I've got tubes and wires running every which way. I looked like a construction site when the electricians and plumbers just walked off the job, you know?"

Apparently he'd had a cardiac arrest. It was a close call, but John had been lucky. He'd had his cardiac arrest in a public place. It turns out that you're much more likely to have someone start CPR if you die in public than you are if you die at home. Even though your family members arguably have a greater incentive to do CPR than a stranger would, in a public setting there are simply more people around, and so it's more likely that someone will take the initiative.

In John's case, two people sitting nearby started CPR, and someone else called 911. Fortunately, the fire station was only two blocks away, so the paramedics arrived in about five minutes. They were able to shock his heart back into a normal rhythm immediately, and they hustled him to the nearest emergency room.

The next day, when a cardiologist offered him an implantable cardiac defibrillator (ICD), John didn't hesitate. He just said, "Sign me up."

Also, John explains, he's divorced and lives alone. He knew there was a good chance that the next time his heart walked off the job there wouldn't be anyone nearby to make that 911 call. An ICD was really his only option.

ICDs have become remarkably sophisticated, but the basic premise is essentially what the name implies. Just like an AED, an ICD has the ability to sense a rhythm, and to shock the heart if that rhythm is abnormal. The basic design consists of wires that are embedded in the muscle of the heart, and those wires lead to a box that analyzes the rhythm and delivers a shock if needed. The box in some of the earlier devices was about the size of a hefty paperback, but they've shrunk over years of evolution to about the size of a pack of playing cards.

There have been advances in the way these devices are designed and implanted, too. Some of the earliest versions required open-chest surgery. That technique gave way to the use of electrodes that were threaded through veins. More recent versions don't touch the heart directly, but rather use electrodes under the skin of the chest.

They are, you have to admit, pretty nifty. For someone like John, who has just realized that his heart is not the dependable buddy he'd always thought it was, these devices seem like a gift. And so patients decide, as John did, that there really isn't much of a choice. "Sign me up," they say.

But not everyone shares John's enthusiasm. Many people refuse the offer of an ICD. Some don't understand the risks of an arrhythmia, and others aren't urged by their physicians to consider it. It doesn't help, either, that these devices have had a somewhat spotty track record of quality issues and recalls. They're still safe, but some people are turned off by stories of devices that fail to work as advertised. In fact, it may be that a majority of patients who could benefit most don't actually get the devices, particularly women and African Americans.

As John and I are talking, the clinic technician knocks on the door and lets himself in. He looks like he's about fourteen—too young to manage something as delicate and as important as an ICD. I'm tempted to ask to see his driver's license, but he and John seem to know each other. John asks the technician about his girlfriend (who is now his fiancée) and the technician asks John about his golf game.

They're chattering away as the technician places a device over John's chest to scan the ICD. In a matter of a minute, the tech has downloaded its history, including a record of the shock it delivered. He can also see the rhythm that convinced it that a shock would be the right thing to do.

"Yup," the technician says. "Sure looks like it fired once." He nods. "Ventricular tachycardia." He nods again, then starts putting his gear away.

I can't help thinking that there's something oddly nonchalant about this interaction. John has just died, and the technician's reaction is: "Well, it looks like you died. Here, let me validate your parking stub. See you in three months."

But perhaps it's not so strange. If you were a patient with asthma, for instance, and you began to feel a little short of breath, you'd use your inhaler. Then you'd feel better. But you wouldn't call your doctor to set up an appointment, would you? No, because that's what your inhaler is for. You use it when you need it, and then you go on with your life.

In the same way, that's how ICDs are meant to work. Granted, their function is a little more spectacular. And although it's possible to take a quick, surreptitious puff on an inhaler when no one's looking, it's a little harder to die and be resurrected without disrupting the flow of a golf game. Although I suppose anyone stuck behind you at the seventh hole could just play through if he's in a hurry.

It's the ability to adopt this attitude of nonchalance that has made these devices so popular. According to one of the largest ICD surveys anyone's ever done, as of 2009, there had been more than 300,000

ICDs implanted. Implantation rates rose in almost all of the countries surveyed, especially in the United States.

But do they work? They do. In one study of patients with heart failure who were at risk of the sort of arrhythmias that ICDs treat, an ICD decreased the risk of death by 23 percent.

That figure gives a sense of the potential benefits of ICDs, as well as their limitations. This reduction in mortality is impressive, particularly if you're one of the people—like John—who is alive because he has an ICD. But 23 percent isn't 100 percent. That is, an ICD isn't going to save the life of everyone who gets one any more than CPR can save the life of everyone who receives it. For starters, as we've seen, an ICD can't treat every abnormal rhythm. Also, people may develop other health problems that could kill them before their heart does. Still, it's possible that ICDs are particularly beneficial for people with very bad heart disease, perhaps because those are the people who are very likely to die of a shockable arrhythmia.

Even though I'm pretty sure I know the answer, I ask John if getting an ICD was the right decision for him.

"No question," he says simply. "I'd do it again"—he pauses—"in a heartbeat." He smiles.

What John doesn't say, at least not immediately, is that although ICDs are effective, they come with a cost. Some of that cost is financial. Estimates are hard to come by, but studies suggest that ICDs cost somewhere between approximately $37,000 and $138,000 per year of life saved. The actual costs depend on the population, how sick people are, and how long they live. Since most of the costs are for implantation, the benefit in terms of years of life saved per dollar spent is greatest for younger, healthier patients. These are patients like John. Looking at these numbers, it's hard not to agree with John's assessment that there was really no question of whether to get an ICD.

The technician is gone now, and we're done. There's nothing John needs to talk to his cardiologist about. The cardiologist will review the results of the ICD's data scan and will be in touch with John if there's anything he should know.

We're walking out of the clinic together, but there is more that John wants to talk about, so we stop at the hospital café down the hall from the clinic. John colonizes a free table and I buy us two cups of coffee.

"You know," he says when I get back, "it's not quite as automatic as they make it seem." He pauses. "I don't want to make it sound like I'm complaining, but it's not a cakewalk."

I'm not sure what he means, but already I'm getting the sense that this is a conversation he wanted to have outside the confines of the clinic. He's not exactly furtive, but it seems as though these are opinions that he'd rather the technician and his cardiologist didn't hear. It's almost as if he thinks he's about to be disloyal.

"When it works, you know, it's amazing. Like a miracle." His ICD, he says, has fired about ten times in the past two years. Several of those times were just like what happened on Sunday. He was standing there, and then all of a sudden, he wasn't. Sometimes he doesn't remember passing out.

But that doesn't mean his ICD is invisible. In fact, he thinks about that ICD every minute of every day. It's a comfort, but a source of anxiety, too.

"You know, it's strange. I'm always grateful when it goes off. Because I know I'm going to be around a little longer. But . . . it changes life in between."

I ask him what he means.

"There's a sort of . . . panic. I even get a little superstitious." He smiles sheepishly. "I mean, I know this sounds crazy, but once when it went off, I'd just walked by this air-conditioning vent in my study. And just a few seconds after I walked by it, I was down on the ground. And I swear, I know it sounds nuts, but ever since, I give that vent a wide berth."

What John is getting at is that while ICDs save lives, they can erode patients' quality of life. Patients with ICDs have a lower quality of life than the general population, and also a lower quality of life than people with pacemakers. Although, interestingly, people with ICDs report a better quality of life compared to people who are only taking medications to control an arrhythmia.

This loss of quality of life seems to be greatest for people like John, whose ICDs are very active. For instance, patients who receive a shock within the first year of having an ICD report increased anxiety, fatigue, and psychological distress compared to those whose ICDs are quiet. Some patients experience symptoms of post-traumatic stress disorder, too.

"That's the tradeoff," John says as he finishes his coffee and stands up to leave. "You live longer, sure. And I'm all for that, believe me. You gain extra years. But you lose minutes and hours along the way worrying about what's next." He shrugs. "But you know it's worth it."

CPR ON THE FRONT LINES

As our four-ton ambulance plunges through light afternoon traffic, scattering smaller vehicles like minnows, the most convincing evidence of the urgency of this run is not our wailing sirens or our flashing lights, or even the near misses at crowded intersections. No, it's the upturned faces that line our route. On either side of us, the pedestrians and drivers offer a mute testimony to the importance of our mission that is more convincing than all of our sound and spectacle.

I don't know where we're going. And I have no idea what we'll find when we get there. But those looks that are directed at us are my assurance that this must be a matter of life and death.

You'd never guess that, though, if you just looked at the two paramedics I'm with. Jason and Garrett have maintained an otherworldly calm since we left their home base. I'm only a "rider" on this trip—an observer—and I'm crammed like a bag of groceries into a small cubby behind them. Yet I'm probably more anxious than they are.

Jason is tall, impossibly thin and long-limbed, and sports a low-maintenance sandy crew cut. He looks like a weekend triathlete—which is exactly what he is—who works as a paramedic to pay the rent and maintain a decent supply of running shoes. But he's brought the energy of an athlete to every call we've been on today, leaping out of the truck first and charging ahead with unrestrained enthusiasm. Now, though, Jason is starting to fidget. He's just started tapping on the dashboard in time to some inaudible rhythm in his head, and is plainly getting on Garrett's nerves.

But Garrett just takes a deep breath every time Jason's palm slaps the dashboard. He's an old soul, apparently. He's also just plain old, at least for this work. A stocky African American man in his fifties, he has graying sideburns and a weekend's stubble on his cheeks. Garrett used to work as an ER aide, he told me, but about ten years ago he decided that he wanted the freedom to work when and how he wanted. He cycled through a variety of jobs until he completed the two years of training required to become a paramedic, and he's been working steadily ever since. (Paramedics are the most advanced category of emergency medical technicians [EMTs], and can perform many of the procedures that you'd undergo in an emergency room.)

The three of us are racing through the suburbs of a small Midwestern city toward an address that Garrett tells me with typical terseness is a "CCRC." He means it's a continuing care retirement community, like the one that Lorraine Bayless was in, made up of independent homes, apartments, assisted-living facilities, and a nursing home. These guys get only bits of information from a dispatcher who is even more pinched with words than Garrett is, but I've learned a little more from Jason. He knows only that there is an eighty-two-year-old female who has been "found down." That is, on the floor.

That description covers a lot of ground, and indeed I suspect it's the ambiguity of this call that is bringing out Jason's tendency to fidget. She may have had a stroke, as Lorraine Bayless did, or a seizure. Maybe her heart has stopped, or maybe she's only fainted. We don't know, and that uncertainty is making him nervous.

The other piece of this story that is unknown is where on the CCRC spectrum this woman is. She could be a healthy, vibrant woman living in her home. Or she could be a nursing home resident with advanced dementia and multiple medical problems. Despite the availability of trained medical staff, nursing home residents who suffer a cardiac arrest don't have a much better chance of survival than someone who collapses on the street does. So this call could end in a miracle, or it could end very badly.

Weaving through traffic that thins as we make a fast right turn onto a smaller street, Jason taps faster and Garrett gives the truck a little more gas, shooting a glance at the monitor that keeps a running log of how long we've been traveling. This will be the measure of our door-to-door response time, and it already reads 8:54. Not bad, but if this eighty-two-year-old female doesn't have a heartbeat, then every second counts. Garrett drags his eyes back to the road, and the truck lurches forward.

When we pull into the CCRC complex, we follow a pleasant tree-lined street that meanders gracefully as if it were paved to follow the course of an ancient river. It winds past redbrick and white-sided bungalows, each with its own driveway and carport. Then we're in front of a three-story apartment building, which I'm guessing is the assisted-living facility, where people with functional limitations or cognitive decline can get extra help.

There is a woman in a nurse's uniform holding a door open for us. She's looking toward us with the same expression with which someone on a life raft might look at the silhouette of a ship that appears on the horizon. She nods vigorously when she sees us. Then she waves, as if to eliminate any doubt, and to ensure we won't drive on past. But she doesn't leave the door to greet us; she actually seems tethered to her post, turning every few seconds to look through the open door behind her.

Caught up in the excitement of the drive, and infected, perhaps, by Jason's nervous energy, I find myself anticipating that Garrett will stop any second now. Jason will hop out and I'll be right behind him. There's no time to lose. We'll be in the building with the defibrillator pack and stretcher in a minute. And we'll save a life.

But this afternoon, what should be a heroic moment of arrival is when things start to go very wrong.

Jason is out the door as we're rolling to a stop, a red nylon backpack slung casually over one shoulder. I'm right behind him until Garrett reaches over and puts a hand on my shoulder. When I turn around impatiently, he tells me, "These calls aren't the easy ones." Unsure what he means, but knowing that Jason is already inside, I nod uncertainly and follow him.

By the time I reach the door, Jason and the nurse have disappeared. I make my way down a short hallway to a door about halfway down on the right, which is hard to miss. There are half a dozen people crowded around, and I assume they're residents.

I follow Jason into a cozy living room whose wide bay window offers a panoramic view of the street outside. There's an overstuffed blue velvet sofa facing the door, and an antique mahogany credenza against the far wall. It's heavily laden with family pictures in a bewildering assortment of frames. There are what appear to be siblings and children and grandchildren, and timeless baby photos that could be of anyone. That's what I notice first.

Another glance reveals a wide bare swath across the top of the credenza. That's when I notice that more portraits—perhaps a dozen—are flung across the floor. A few frames are still intact, but most have broken and there are shards of glass glinting wickedly in the sunlight.

A second later, I can see that the cast of a hundred or so family members still on the credenza is looking down on a strange scene in which Jason is bent over a very pale, thin woman in a flowered housedress who is lying flat on her back. She's wearing threadbare green ankle-high slippers with pointy toes that seem to tower over her.

There is a young, clean-cut man doing chest compressions that I can't help noticing are too shallow and much too slow. He stops, and Jason kneels and checks for a pulse and breathing. And somewhere in the flurry of efficient activity, as he's taking the defibrillator out of his bag, he finds a second to catch my eye and shake his head, pointing his chin toward the man next to me. He's also in his eighties, with a narrow face and a wild shock of white hair that hovers over his creased forehead.

Jason's message is clear: No heartbeat. No respiration. Take the husband and get him out of the way.

It doesn't take more than a moment to reconstruct the events that have brought us here. An elderly woman, in the middle of dusting the family portrait gallery, feels faint. Maybe she has chest pain, or maybe she's just lightheaded. In her panic, she reaches out in search of a handhold, but her frail fingers find only the portraits. Her flailing hands sweep across the credenza's surface, bringing a dozen of her family members with her as she slumps to the floor.

Imagining that scene, or something like it, I take the stunned man by the elbow and we move over to the window. As much to distract him as in hopes of gleaning useful information, I ask him what happened. His name is Richard, he says, and he begins to explain that the woman on the floor is his wife, and that they've been married for sixty-one years. He repeats this fact several times, as if somehow that number will serve as a charm, holding any bad luck at bay.

Her name is Florence, he says. Then, a little louder so Jason can hear him, he emphasizes that it's Florence, not Flo. She hates being called Flo, he says, more softly. But Florence isn't hearing anything.

In a flurry of activity, the young man doing CPR has stood up and stepped away. Garrett has joined Jason with another red nylon backpack, which he seems to set down and open simultaneously. Then everything happens too fast for me to follow, even though I know what I'm looking for. Garrett has opened Florence's housedress, baring her bony chest while Jason continues CPR. Now there's an oxygen mask in place and Garrett's ropy forearm is bulging as he squeezes the bag with a hubcap-size hand to inflate her lungs.

Garrett turns the defibrillator on, and from my vantage point by the window I can see the jagged white line of ventricular fibrillation racing across the small screen. That line is an urgent call for someone to apply the defibrillator paddles, but they don't do this. Instead, they continue doing chest compressions, and delivering breaths through the oxygen mask.

Jason and Garrett know what most impatient bystanders don't—that CPR by a paramedic is as good as an immediate shock from a defibrillator. They're being careful and methodical. I know this too, and yet as the seconds tick by I start to worry. What are they doing? I'm increasingly feeling the urge to leave Richard's side and pick up the defibrillator pack myself.

Soon, though, just as I knew he would, Garrett applies one shock, followed by a second. With each jolt, Florence's limbs move and her chest rebounds as if a small animal is trapped inside her chest and is butting its way out. One of them—Jason?—has started an IV in Florence's left arm. There's a third shock, and then a fourth. They push saline solution through the IV, then a dose of epinephrine, then lidocaine.

As this drama unfolds, I'm trying to get a little more information about Florence, which I pass on to Garrett and Jason as they work. She has severe heart failure, Richard says, for which she's taking a long list of medications. She has diabetes, too. Her heart failure has been getting much worse, and she's been in the hospital three times in the past three months. And she has liver disease. But it's not because of drinking, Richard hastens to add. It's something to do with her heart.

Out of the corner of my eye, it seems as though the tempo of Jason and Garrett's efforts has slackened just a little. They both sit back for a second and I can see the monitor showing something like a regular rhythm. Glancing up, Jason confirms with a curt nod that Florence's heart is back in sinus. She has a normal rhythm, and his nod to Garrett, who has two fingers on Florence's neck, confirms that she also has a pulse. Without saying another word, they both seem to come to a consensus: it's time to go.

In what seems like only a few seconds, he and Garrett have Florence positioned on a stretcher. The defibrillator is nestled on top of her. Now Jason is squeezing the ventilator bag, manually pumping oxygen into her lungs. Neither of them looks at me as they pass.

In the county morgue where I did a rotation as a medical student, I met a technician who used to be a medic in the Marine Corps. One of his morgue duties was to sew up bodies after an autopsy, and he performed that task with a speed, efficiency, and economy of movement that was truly striking. In no more than a minute, he'd restore the contours of a body that the pathologist had torn apart, readying it for the undertaker. He was so quick and sure that even the pathologists would stop whatever they were doing to watch him. But what I remember most vividly is that despite his technical prowess, he never revealed any pride, pleasure, or satisfaction in his work. In case after case, he would sew, knot, and cut the heavy nylon thread with the same blank, vacant expression.

I saw that same look on Jason's and Garrett's faces as they wheeled Florence out the front door. Although they'd achieved something undeniably remarkable—they'd brought a woman back to life—there was no pride and no excitement. No more than there was for that ex-medic sewing up a patient whose days were over.

Five minutes later, Garrett and Jason have loaded Florence into the ambulance, and we're racing toward the nearest hospital. Garrett is driving and Jason is in the back with me, tending to Florence. Through the back window, I can see the suburbs flashing by, and although I can't tell where we are, I count three turns in quick succession.

Whenever Garrett slows suddenly for those turns, Florence slides forward just a bit and her head bumps the stretcher's metal railing with a soft thud. It's a little thing, not enough to cause any discomfort if she were awake. But each time, those bumps offer a monotonously gloomy counterpoint to the monitor's optimistic chirping.

Soon we stop abruptly and back up to the emergency room's trauma bay. When the ambulance doors open, the quiet of our little world splinters and Florence's stretcher is whisked out by a dozen strong hands. Now this is for real. There's no room for observers, regardless of what letters I might have after my name. My only job is to get out of the way.

Florence is handled with a cool efficiency, and a team of blue-scrubbed staff wheel her at a controlled jog through the oversize sliding doors and into one of two trauma bays. A nurse is performing chest compressions as they move. I follow cautiously to find the team placing a large IV line in a vein in her neck, a catheter in her bladder, and an endotracheal tube down her throat. A lead-aproned X-ray technician is hunting for space around the table, and jockeying for a few seconds in which to snap a chest image. And all around Florence, there are tubes and wires and monitors and eight or nine people all focused intently on bringing her back to life, one more time.

There is nothing in this scene that's new to me. I've done everything that the team is doing. I've run codes, and I've applied defibrillator paddles and placed intravenous lines in the jugular vein.

But this is the first time I've seen those things done on someone I've "met," whose image from an hour ago is as vivid in my memory as the scene playing out right now is. Although I've secured an endotracheal tube in place with the same ubiquitous fabric tape that the respiratory technician is applying diligently, now I'm watching the process with a clear image in my mind of Florence's face as it had been only moments before, free and unencumbered. As that tape obscures Florence's sharp cheekbones, I'm thinking of the dozens of photographs on the credenza back at the CCRC that display the same vaguely Nordic family resemblance. Although I anticipate Florence's galvanic hand-twitch in response to a 200-joule shock from the defibrillator paddles, it's strange indeed to visualize those hands an hour ago, busy rearranging her family photo gallery.

Half an hour later, Florence is stable. She has a normal heart rate and her blood pressure is OK. She's not awake, but that's because she's been heavily sedated. She's still on a ventilator that is breathing for her, so it's too soon to declare a victory. It's too soon, even, to try to predict whether she'll ever wake up. But she's stable enough to transfer from the ER up to the ICU.

Word comes in that Richard is in the waiting room, with their daughter. I go out first, as the team is finishing up, and the family resemblance is strong enough that I recognize the woman at the registration desk—tall, blond, in a crisply tailored business suit—as Richard and Florence's daughter, Carla. I introduce myself, and we sit down to talk until the ER physician can come out.

Almost immediately, Richard asks me how Florence is. I start to explain that her doctor is on his way out, but he interrupts me, asking whether she's alive. That's all he wants to know, he says. I wonder, Can I tell him that much?

I tell them that she is. Yes. That much I can say.

Richard and Carla's response is oddly mixed. There's relief, of course. I can see it plainly on both their faces. But it's relief tempered by the realization that if Florence is alive, then the real roller-coaster ride is just beginning.

The physician comes out then and introduces himself. We all sit down and he tells Richard and Carla what he knows at this point. But that's not much, he's quick to clarify. It's far too soon to say anything definitive about how Florence is doing, or—he pauses—about how things will turn out.

First he tells them a few things about which he's certain. For instance, he says that for the past twenty minutes in the emergency department her heart's been in a stable rhythm. Still, she's not breathing on her own, he cautions them, so she's on a breathing machine. And she's unconscious.

It's clear to me from the facts that the physician has selected, and from his steady tone, that he's hanging crepe. He's giving Richard and Carla all the signals he can that Florence is in very serious condition. And he's warning them—subtly, and without making a formal prediction—to prepare for the worst.

Then Carla asks point-blank whether Florence is going to live, and the physician frowns. It's a difficult question. He thinks for a moment, and then answers that every hour she's alive, her chances are a little better. Take one day at a time, he says finally, laying one hand on Richard's shoulder.

But he's not describing Florence's chances for a total recovery. The most basic fact is that fewer than 10 percent of patients who suffer a cardiac arrest outside a hospital survive to leave the hospital. Based on that fact alone, Florence's chances are slim.

That's an oversimplification, though, which glosses over numerous factors. Indeed, the equation that will determine Florence's survival is much more complex than a simple number would suggest.

Florence has other factors in her favor. For instance, Jason and Garrett arrived quickly, and they worked fast. So did the young man who started CPR. (I learned later that he was an activities coordinator—ironically the same position that Colleen held at Glenwood Gardens, where Lorraine Bayless lived.) Also, it's good that Garrett and Jason were able to restore a heartbeat before they transported her. She's also at the best hospital in the region.

The conversation is beginning to wrap up when Carla asks whether, if Florence wakes up, her brain might be . . . damaged?

The physician pauses only a split second before retreating to the only answer he can offer.

"It's too soon to tell," he says. "Let's try to get her through tonight, then the next couple of days. Then we'll see."

Garrett looks up as I open the door to the break room, where he and Jason are finishing the paperwork for their shift. He seems tired. Tired and about ten years older than he'd been when I met him that morning.

Although I have some questions for them, Garrett's haggard look reminds me that they're officially off duty. It's seven p.m. and they've just finished a grueling twelve-hour shift. They probably want to get their paperwork done so they can leave, so I figure I'll just thank them for letting me hitch a ride.

But Garrett looks up and smiles. Then Jason kicks a chair out from under the scuffed Formica table that is strewn with spreadsheets and checklists. He gestures for me to sit, and I do. I'm wondering what's on their minds. I don't have to wait long.

Garrett flicks his pen onto the table, where it lands on their shift call list, spinning crazily in an arc that flicks among the diagnoses we've seen that day like some morbid game of spin-the-bottle.

"Motor vehicle accident."

"Bicycle accident."

"Fall."

And, of course, "Cardiac arrest."

Garrett breaks the silence by asking me what I thought of Florence's case.

I say I'm not sure what he means.

"What do you think about how it turned out?" he asks. "And what we did?"

I'm still not sure what they're probing for, but I'm beginning to get an idea. So instead of answering, I ask how *they* think it turned out. They see these cases all the time, I point out. They know better than I do what the possibilities are. What did they think? And then, because it seems like a little extra prompting might help, I ask them if they think they did anything wrong. That, it turns out, is exactly what they need to start talking.

Garrett goes first, wading in with the studied intensity of someone who, I'm guessing, has spent a long time—and a lot of futile calls—thinking about what he was doing, and why.

"We have to try," he says. But as he says that, he's also shaking his head with a strange intensity. It seems as though he's trying to reconcile an internal disagreement.

"That's the thing," he continues. "When we're there, whether it's a kid who's choked on something, or an old lady with dementia and cancer in a nursing home, we have to do everything. Doesn't matter. When we first walk in that door—hell, when the call comes in, mostly, we know what's going to happen. It's not like it's a mystery. The choking kid? No problem. He'll be fine. But the lady in the nursing home? No way. Never. We know when we get the call that it's going to be a waste of time. But we have to do it. We have to go. And we have to go *fast*."

He's not shaking his head anymore, and I can't help thinking that, somehow, in that short explanation, he's reached some sort of internal compromise.

Jason has a slightly different take. He's been tapping his fingers furiously on the tabletop for the past thirty seconds, like he's waiting to have his say. But as he starts speaking, he's not voicing an opinion so much as he's trying out an idea. Thinking out loud.

"Really," he says, "people don't want this, you know?" He looks at Garrett, who shrugs. Then he looks at me. "A lot of people don't, anyway. They don't want some stranger storming into their home, pounding on their chest, and doing all sorts of invasive things to them while their family is watching. They don't want to be thrown in the back of a truck and carted to a hospital. And they sure as hell don't want to die in an ICU, unconscious and connected to all those tubes and wires."

He turns to Garrett. "Remember that lady back about two months ago? The tomato lady?" Garrett nods and Jason turns back to me. "Same neighborhood as the call we just did. But a private home—big house. Older lady, lots of health problems. Collapsed in her garden one morning—she was out watering her tomatoes. So her husband called us. We did what we could and we brought her back, but she died in the ER. And you know what her husband said to me? He said he felt awful, because he was certain that if she had to die, she would have wanted to die right where she was, in her garden."

Jason shakes his head. "You see what I'm saying? A lot of the time we're not actually letting people live any longer, we're just changing how they die."

"And we're giving them all kinds of hospital bills up along the way," Garrett adds.

I ask them about so-called out-of-hospital do not resuscitate (DNR) orders. The order is actually a form—like a living will—that a patient and doctor fill out together. It's usually accompanied by a bracelet that tells EMTs and bystanders not to do CPR. But Garrett says they never see these forms. Do they see bracelets often? I ask.

"No one wants to think about these things, you know?" Jason says. "I mean, you don't think about it until you have to. . . ."

"And then it's too late," Garrett finishes the thought.

I think that's a pretty good summary about the way we make most end-of-life decisions. We avoid making any decision until we have to, and often at that point we're too sick to make a decision at all. Then our families have to make a decision for us.

"But there's no alternative," Jason says. "Either someone says they don't want to be resuscitated, or they'll have us working on them. Those are your only choices."

Jason seems frustrated by this, but Garrett is more philosophical. "That's just the way it is," he says. "We can't make the call. And I for one don't want to. Wouldn't want to. I'd quit before I took on that kind of responsibility. Make a decision in the field about whether to resuscitate someone? No way."

Jason is more open to the idea. "I wouldn't like having to make that call either, but I'd do it if I could. And for sure I'd listen to the family, if there's time. At least we should be able to give them a chance to call it off. If the family says stop . . ."

"But that tomato lady's husband wouldn't have told us to stop." Garrett raises the obvious objection. "And today? Her husband? Never. Even if we could ask them, most families wouldn't."

That's probably true. Especially when the paramedics are already there, in the house. Let's face it, Jason and Garrett look like pros. They look like they know what they're doing, and they inspire confidence. It would take a strong personality to tell them to stop.

And Garrett and Jason and thousands of paramedics like them can't make the decision to stop. They operate under a central medical control, with strict protocols to follow. They're obligated to attempt resuscitation on all patients, unless there is clear evidence, like rigor mortis, that a patient has been dead for an extended period of time.

"Families don't want to give up," Garrett says. "Even if a patient's ready to go, that doesn't mean the family's ready to let him. They'll call 911 and then cut the bracelet off."

I ask if that's ever happened and Garrett and Jason shrug in unison.

"Hell," Jason says. "Maybe it just happened. We'll never know."

Several days later, I visit Florence in the ICU, where she is still unconscious and on a ventilator. I notice a picture of a much younger Florence taped to the wall above her head. Dressed in white, and wearing a complicated hat, she looks like she might have been at a summer wedding. Carla told me later that it had been taken at Carla's wedding, twenty years ago.

Richard says they put that picture up to remind the doctors and nurses of who Florence had been, and who she really was. Carla adds that it's partly for them, too. "I don't want my last memory of her to be in this bed, like this," she tells me. "But I hope we did the right thing," she says, reaching over to squeeze her father's hand.

Richard did the only thing he could have done. He picked up the phone and dialed 911. Maybe it was inevitable.

There are many costs to what Florence and her family went through, but Carla is wrestling with one that we tend to ignore. The lingering uncertainty that is going to haunt them for years. Maybe Richard shouldn't have called 911. Or maybe he and Carla shouldn't have agreed so readily to aggressive treatment once she reached the ER. Each time the question arises, they'll feel guilty for doing too much.

They'll also have to live with the memories of Florence as she is now. No matter how intently they focus on it, that picture above her bed isn't going to eclipse the very real Florence right in front of them, with a tube down her throat and the wheeze of the ventilator in the background. Just like their feelings of guilt about maybe making the "wrong" decisions, that image will be with them for a long, long time.

Finally, there are the financial costs of Florence's care. Conservatively speaking, the price tag for the ambulance ride, the ER treatment, and a week in an ICU is probably more than $100,000. Most of those costs will be paid by Medicare, but they are costs nonetheless.

And they incurred all these costs—financial and psychological—for . . . what? Florence so far hasn't had any additional meaningful time with Carla or Richard or her grandchildren. She has yet to do, or say, anything. All of that treatment hasn't bought her any real time. Sitting here in the ICU with Carla and Richard, I'm thinking that it's as if Florence really had died a week ago but is condemned to live in some half-dead state until her family lets her go.

A week after I saw them for the last time, I heard that Richard and Carla decided that enough was enough, and that they should stop aggressive treatment. They had different reasons, apparently. Richard had decided that Florence wouldn't survive. On the other hand, Carla was worried that Florence would survive, but in a state that she would have found unacceptable. From those very different fears, the two of them came to a shared decision to let her go.

So the medical team turned off the ventilator and removed the tubes and wires that had been keeping Florence alive. She breathed on her own for a few brief minutes. Then her heart rhythm faltered and her heart stopped for the last time.

7

When Is "Dead" Really Dead? Listen for the Violins.

"DEAD" VERSUS SINCERELY DEAD

I set out a year ago to learn about the science of resuscitation from the other side of the fence. That is, from the patient side instead of the doctor side. When I began, I'd seen many of my own patients—too many—reach the end of the proverbial road. These were people whose hearts had stopped working permanently.

Some of those people had been through hell, like Joe, one of the first patients I met as a medical student and to whom you were introduced in chapter 1. He spent eighteen unconscious days in an ICU thanks to my efforts at CPR, and his family endured eighteen days of miserable watching and waiting because of me. I looked back at their histories, and at what they and their families had been through, and I became convinced that much of what they endured was futile.

But there were others—people like Michelle Funk—who were alive against all odds. These were people who had bounced back after their hearts stopped. In short, people who had benefited from a miracle.

I wondered what the future for all of us is going to be like. Are we all destined to die the way that Joe did, slowly, and in stages? Or can we look forward to miracles like the one that saved Michelle's life?

Those were the questions I started with a year ago. So what have I learned? Well, first, I've learned that death isn't what it used to be.

A man is shot multiple times and lies dying on the ground. His girlfriend is at his side to comfort him in his last minutes, and they say their tearful farewells. Weeping violins fade in. Then a bystander breaks the spell by asking how the man is doing. The girlfriend, annoyed, replies: "He's dead! Can't you hear the music?"

Some version of that scene (in this case from the spoof *I'm Gonna Git You Sucka*) has been reprised countless times on stage and screen. A character dies and there are vivid, unmistakable cues. Most obviously, there's the swell of music in the background (Samuel Barber's "Adagio for Strings" is a favorite: see, for example, *Platoon*).

But there isn't always an orchestra on call for these events. In that case, we need to rely on someone who can announce the person's death. Arguably the most famous authority in that role is *Star Trek*'s Dr. Leonard McCoy, who seems to relish the opportunity to announce, "He's dead, Jim."

If Dr. McCoy is off in another galaxy and can't be reached for comment, there are other cues that are subtler but still unmistakable. The victim's eyes stare blankly (*Dead Man Walking*) or, for a more high-tech cue, the lights in Dr. Octavius's arms die out (*Spider-Man 2*). Or a hand goes limp and an object falls to the floor (a snow globe in *Citizen Kane*). Or, in what has to be the most unmistakable death scene in the history of cinema, when Jimmy Durante's character dies in *It's a Mad, Mad, Mad, Mad World*, he kicks a bucket. Literally.

Yet the paradigmatic example of this trope, at least in my book, occurs in *The Wizard of Oz.* The Munchkin coroner leaves no room for uncertainty in viewers' minds regarding the fate of the Wicked Witch of the East. He proclaims:

"As Coroner, I must aver
I've thoroughly examined her
And she's not only merely dead
She's really most sincerely dead."

And there you have it. Not just dead. But sincerely dead. Which is pretty darn dead.

In all of these scenes, death is unmistakable. Moreover, the point at which death occurs is unmistakable. There is a clear, bright line. And once a character crosses that line, there's a cue for lights and music and McCoy's famous line.

A hundred years ago, maybe, those scenes might have been a pretty good representation of how deaths played out. Someone was alive and then—suddenly, clearly, and unmistakably—they weren't. Bystanders noted this fact and then moved on. Maybe sad music played in the background. But even without the score of "Adagio for Strings" playing in their heads, people knew that the person in front of them was dead.

But now? Now, it's not that simple. Now that line between alive and dead is getting increasingly blurry precisely because of the sorts of advances we've seen.

Remember Michelle Funk? She was lucky that her rescuers were blissfully deaf to the sound of violins. But she's hardly the only person we've met who was saved because no one stopped to listen. What about Anna Bågenholm, the "ice woman" who survived for hours? Or Mitsutaka Uchikoshi, the human bear who figured out all by himself how to hibernate? Or Thomas, the mushroom farm worker whose brain was put on ice for surgery? Or the dozens of bodies (and heads) stacked in that Alcor warehouse in Scottsdale, all of whom have completed their first life spans but who, we're told, will be back? Someday.

Each one of these examples has blurred the line between life and death a little more. The result is that it's often no longer possible to say with the certainty of a Munchkin coroner that someone is "sincerely dead." Because as soon as you do, a second later you'll think about those people like Michelle Funk or Anna Bågenholm who were also "sincerely dead" until, all of a sudden, they weren't.

WHAT'S POSSIBLE?

If the line between life and death is blurry now, what is it going to look like in five years? Or ten? Or fifty?

From everything I've learned in the past year, I'm betting that the science we've seen up until now is just the warm-up act. I think the pace of advances is going to pick up, and I wouldn't be surprised to see the field grow exponentially over the next ten years. A year immersed in the science of resuscitation hardly qualifies me to predict the future, but I'll offer a couple of guesses.

First, I'm betting that we can look forward to a huge bump in survival rates after cardiac arrest. How huge? Probably in the next five years, patients will be able to expect a better than 50 percent chance of a good neurologic outcome. Overall survival, too, will go way up. That's partly because of the science I've seen in my travels, of course. But also because of relatively low-tech innovations like AEDs on every corner. They're not fancy, and the technology that makes them possible has been around for decades. But they save lives, and the more of them are out there, the more lives will be saved.

That's the next five or ten years. But what can we expect in twenty or thirty? Once we learn how to bring people back to life, and once we get good at it, what's the next big thing?

Honestly—and I'm going out on a limb here—I'm optimistic about suspended animation. Not the crazy space travel version. But short-term suspensions that keep people alive for hours or maybe days.

I'm not taking any sides in the squirrel-lemur showdown, and I'm open to the possibility that there are other examples out there that we don't know about yet. Maybe that's Dr. Cheng's AMP, or maybe it's something else entirely. But someone, somewhere, is going to figure out how to place people in a state of reduced metabolism that's a small fraction of normal. And when they do, that's going to be huge.

HAPPY FAILURES

If it sounds like I'm optimistic, I am. In part, my optimism comes from the rapid progress that has been made in the past ten or twenty years. That is, I'm optimistic because of our successes.

But I'm also optimistic because of our failures. And there have been a lot of those, most of which I've left out of this book. For every day that I spent with people like Cheng Chi Lee and his mouse #0011 that entered a state of suspended animation, I spent weeks tracking down advances that didn't pan out.

Hydrogen sulfide (H_2S), for instance, was the next big thing, until, all of a sudden, it wasn't. H_2S was first studied in detail by a biologist named Mark Roth, gaining him instant fame and, not coincidentally, a MacArthur "genius" award. Apparently this stuff—the gas that's emitted from rotten eggs and sewage-treatment plants—induces a state of hibernation in mice. This would be the new wonder drug, everyone thought.

In an experiment that received a blizzard of publicity, Roth exposed mice for six hours to H_2S at a concentration of 80 parts per million. His team observed an average drop in temperature of 13 degrees Celsius and a 90 percent drop in metabolism. Impressive.

Roth's results inspired a wave of follow-up studies, but those studies, alas, produced results that were not so inspiring. Pigs, apparently, don't respond very well to H_2S. Nor do sheep.

However, humans are not sheep, or pigs. So, undeterred, Roth brought H_2S to clinical trials in people. That created challenges, because H_2S is a gas, and you can't use gas safely without the risk of poisoning everyone nearby. So Roth needed a form of H_2S that could be administered intravenously.

An early study used sodium sulfide (Na_2S). Exposed to water and oxygen (for example, in humid air or in the bloodstream), Na_2S generates H_2S. It's not without risks, though. Na_2S is strongly alkaline in solution, like sodium hydroxide, also known as Drano. OK, so this isn't sounding like something you want to get anywhere near, right?

And indeed it turns out that you don't. That trial began in May 2009 and was terminated in April 2010 after only six subjects had been enrolled. As is often the case when trials are stopped, available information is sparse. A query to Ikaria, Roth's company, received only this terse reply: "Unfortunately, I'm not able to provide any further details other than our development of this asset has been terminated."

So hydrogen sulfide, apparently, is not coming to an emergency room near you anytime soon. And who knows? Maybe Cheng's AMP will suffer the same fate. Indeed, it seems like many innovations that look exciting now will be doomed.

Failures like this are cause for optimism because they mean that science is taking risks. People are asking questions that have a very small chance of turning up answers that will lead to new treatments. And when there are a lot of scientists trying new things—and taking nosedives—it's a comforting indication that we're exploring new avenues and that, eventually, we'll learn something.

And the reason these risks are possible is the last reason I'm optimistic. Over the past year I've spent nosing around and talking to scientists, I've been amazed at how many receive funding from the private sector. I'm even more amazed that many have started their own companies. And these aren't businesspeople. They're scientists, backed by venture capital. So someone out there—many someones, with hundreds of millions of dollars to spend—is betting that resuscitation science is a good investment opportunity. And if you follow the money, it's safe to say that the science of resuscitation is going places.

SO YOU WANT TO BE A CRYONAUT

If the science of resuscitation is going to be increasingly characterized by risk taking and frequent failures, then some of the ideas I've seen in the past year are going to end up in the wastebasket of science. But which ones? Well, at the top of my list of likely failures, at least in our lifetimes, is cryonics.

The biggest challenge facing cryonics is the enormous gap that exists between what can be done in a laboratory and what cryonauts are currently attempting on people. In the laboratory, we've gotten to the point where we can freeze eggs and sperm and bits of tissue, like corneas and heart valves. What cryonauts are doing—with fingers crossed—is freezing whole people or sometimes just their heads. Contrast that with the pace of other advances we've seen, which have moved carefully, methodically, and with many stops and starts, from animals to people. The work on suspended animation, for instance, has been slow but steady.

The problem isn't only that cryonauts are trying things in people that haven't been tried in animals. The real issue is that there's none of the usual back-and-forth of failures and successes that will help us learn. There have been no whole animal successes at all yet, anywhere. There's no process of perfecting a technique in animals and trying it out in humans—and failing—so we can go back to the drawing board. Hydrogen sulfide was both a miracle and a bust in the space of ten years. We learned something, and then science picked itself up and moved on.

But in cryonics, they're freezing cryonauts first, and hoping that science catches up, not just in terms of thawing out the cryonauts, but also in curing what was killing them in the first place. It's a little like tickling dead people with feathers and hoping that, someday, science will come along and prove that feathers really are a good strategy for life preservation.

Then there's the quackery quotient. For the most part, and with a few exceptions, the science just isn't there. Most of the work in cryonics is self-funded or supported by small foundations. Little of it is peer-reviewed, and hardly any ends up in mainstream scientific publications.

Of course I understand that mainstream science is often risk-averse and resistant to new ideas. And I'll also admit that some of the most impressive advances in other areas of science have been made without following the usual pathways of funding that flows from the National Institutes of Health to top-tier research universities. Craig Venter's effort to sequence the human genome, for instance, was a private end-run around mainstream science.

But Venter was a respected scientist who used techniques that were widely accepted. Moreover, the premise of his work—sequencing the human genome—had already gained wide support. So Venter's foray into private science was really just a way of advancing the goals of mainstream science more quickly and efficiently. Cryonics, on the other hand, represents an entirely new direction based on new ideas and assumptions.

I'm also pessimistic about the future of cryonics because most science—good science—is a numbers game. Advances are usually made by making mistakes. Lots of them. That's multiple attempts over years, and multiple failures, in hopes of getting to something that is not a complete failure. And then hundreds more trials in order to get to something that someone would call a success.

That's not true all the time, of course. But if the stories of scientific advances in these pages are any indication, it's certainly the norm. Think, for instance, of all of the rodents, rhesus monkeys, and "mongrel dogs of medium size" that have given their lives for the pursuit of science. And although I'd be the first to admit that we've made a lot of progress, it's still the case that progress in terms of survival after a cardiac arrest is measured in minutes.

You could say, of course, that the science of cryonics could be advanced with animals. And indeed it could. (Although I suspect if they could, vast swaths of the animal community, including lemurs, dogs, rhesus monkeys, mice, pigs, and of course thirteen-liners would all shout, "Not it!")

The problem is that whereas dog hearts work pretty much the way that human hearts do, dog *brains* are, sad to say, different. Dogs don't do many of the things that people do, like admire a sunset, microwave popcorn, or splash in the pool with their kids. So it's difficult indeed to guess at what, if anything, a thawed golden retriever's cognitive status will tell us about a thawed cryonaut's likely ability to do any of these things in a thousand years.

If you want your future to include sunsets, I'm afraid cryonics may not be your answer. There's just not enough serious science there to produce results that any of us are likely to enjoy. So my advice is not to count on a second life in a thousand years. Watch a sunset now. And—why not?—make some popcorn while you're at it.

"MORAL HAZARDS" AND THE PRICE OF RESURRECTION

Absent from much of this discussion of the scientific horizons so far is any mention of costs. But those costs are very, very real. Remember, we're talking about the costs of resuscitation, as well as the costs of hospitalization afterward. That can run more than $20,000 per day. Add the costs of implantable defibrillators, and you're into the six-figure range.

Finally, there is the cost of the procedures—hip replacements, heart bypass surgery—that would have been rejected as too risky in many patients twenty years ago but are now routine for almost everyone. Those price tags are just a tiny fraction of the total costs of resurrection technology. For a true reckoning, you'd need to consider the cost of a hip replacement for a woman who wouldn't have been an operative candidate ten years ago. Or a man who is now eligible for dialysis because we're confident that we'd be able to resuscitate him if his heart stops during a dialysis session. A patient who would have been too sick for chemotherapy, surgery, or dialysis in the past now becomes a patient who is merely "high-risk."

This is one of the hidden costs of the advances we've seen in the science of resuscitation. The ability to bring someone back to life provides a kind of safety net that makes certain treatments possible. With that safety net in place, doctors are able to be more aggressive with their recommendations for treatment than they would otherwise be. They can push the envelope because they know that, if a patient really gets into trouble on the operating table, they'll probably be able to save her.

This safety net effect is pivotal in the way that doctors make decisions, because doctors are generally risk-avoidant. We don't want to use a medication or perform a procedure if there's a chance—even a small chance—that doing so will result in a patient's death. The Hippocratic Oath's admonition to do no harm is in the back of our minds whenever we recommend a treatment or write a prescription.

But what if that harm could be avoided, or mitigated substantially? What if we knew that, if the worst happened, we could bring a patient back?

In other settings, this phenomenon is often described as a "moral hazard," because a safety net lets people take risks that they wouldn't ordinarily take. Think for a moment about the way that bankruptcy laws work. If you get into serious financial trouble, from which you can't dig yourself out, there's the option of declaring bankruptcy. That may leave you with little or nothing, and will wreck your credit for a long time, but at least you can erase your debts and start over.

That seems like a small thing, perhaps. Although clearly it's not so small if you're the one who is clawing back from a ruined financial life. But it's actually a dramatic change from a rough-and-tumble history in which debts were punishable by slavery (ancient Greece), imprisonment (nineteenth-century England), or even death (Genghis Khan's mandate, at least for repeat offenders).

Bankruptcy is no picnic. But it beats the heck out of slavery. So it stands to reason that we might be just a little more willing to ask for a loan from Aunt Mabel if we can be sure that we won't end up in chains in her dungeon.

Just as bankruptcy laws and insurance may encourage riskier behavior, the availability of resuscitation creates a moral hazard too. Treatments that would have been "too risky" only twenty years ago because of the risk of cardiac arrest are now routine. Many of my oncology colleagues tell me they've become more aggressive in using chemotherapy in older, sicker patients too.

The most fascinating thing about this safety net is that there's a good chance that most of us have been affected by it. If you've undergone any procedure as an outpatient, for instance, that procedure was possible because of advances in life-saving technology. Procedures like wisdom tooth extractions or endoscopy or even hernia repairs that used to be conducted in the operating room can now be conducted in an outpatient surgical suite.

Why? Because the science of resuscitation (even in a limited outpatient setting) is light-years ahead of what could have been accomplished by an entire team in an operating room fifty years ago. Back then, if your heart stopped during a procedure, the best course of action involved cutting you open and applying wires directly to your heart. If you've had your wisdom teeth taken out, I can guarantee you that an open-chest resuscitation wasn't on the menu if something went awry. Of course, the shift to outpatient procedures has been pushed along, too, by improvements to technology that make procedures less invasive than they once were. But none of those advances would have been possible if advances in resuscitation hadn't kept pace.

Taking all of those costs into account, it's safe to say that the advances in the science of resuscitation that we've seen already come at an enormous cost. That's not to mention whatever new advances are on the horizon, such as cooling techniques like the RhinoChill, deep hypothermic surgery, or—just for the sake of completeness—cryonics. Each one comes with a price tag.

HOW MUCH IS TOO MUCH?

If it sounds churlish to talk so bluntly about the costs of being alive after a successful resuscitation, it is. It is churlish and brutal and callous. And so we never talk about these costs. Not at the bedside, not at ethics conferences, and certainly not in public policy circles.

I met a man once, a friend of a friend, who had survived a harrowing resuscitation after a cardiac arrest. He had every complication imaginable, and a few that his doctors had never seen before. When he finally emerged from the hospital after four months, he faced a total hospital bill upward of $5 million.

On a hot, lazy summer afternoon, we sat on his deck talking about his experience. We talked about what he knew, and what he was told. And we talked about what his life had been like since. Our conversation meandered for more than two hours. But never in all that time did we ever talk in detail about the costs of what he went through, and whether his survival was worth it.

I can't imagine questioning the price of technology that allowed him to be sitting on his deck, sipping iced tea, and talking to me as his grandkids played Marco Polo in the pool behind him. I would have to be coldhearted indeed to sit there and think—much less *say*—that he shouldn't be alive. Or even that there must be a point at which his being alive was just too expensive.

Instead, we ignore those costs when we're making decisions about treatment. We pretend they don't exist. We talk instead about quality of life, and survival. We sometimes resort to other euphemisms, like a nod to a scarcity of ICU beds. But questioning the costs of saving a life is strictly off-limits. Of course, weeks or months later, patients and families facing bankruptcy have to talk about costs. But that's too late.

We ignore costs as a factor in decision making because the stakes are just too high, and the possibility of a mistake is just too great. Just as Michelle Funk's rescuers did by the side of a swollen creek, someone will think that maybe this is a life that they can save. It's a long shot. Maybe a very long shot. But you never know, so . . .

That's why I'm pretty confident no one is going to impose serious limits to technology anytime soon. As new advances become available, we'll reach for them. In hospitals, on ambulances, and in shopping malls. Better technology that offers even a tiny chance of saving a life will be embraced.

Conversely, I can't see any future in which there are widespread decisions to limit or withhold that technology. Remember the uproar that ensued when Lorraine Bayless wasn't resuscitated? She was a woman for whom CPR probably wouldn't have been successful. Even if it were, "success" would have meant waking up to a severe stroke. And, finally, remember that her family said she had wanted to die a peaceful death. And yet, when CPR was withheld, it resulted in a national incident. So it seems much more likely that new technology will be deployed as it becomes available.

That doesn't mean there won't be rationing. Any new lifesaving technology almost certainly will not be evenly distributed, and it's likely we'll continue to see AEDs in suburban shopping malls long before they appear in inner-city bus stations. And people who live in communities with a strong tax base will stand a better chance of a successful resuscitation by first responders. That sort of rationing may not be as obvious as a hard-and-fast rule about who should be resuscitated would be, but it's just as real.

MICHELLE FUNK—THE MIRACLE GIRL

Isn't it worth it? Isn't saving a life worth whatever it costs? If anyone could be the test case for that proposition, it would be Michelle Funk, the miracle girl who survived for three hours without a heartbeat. Can we count her story, at least, as a major victory?

Indeed we can. I'm delighted to report that she's alive and well. She just got married in 2012, in fact. (She and her husband, Michael, were registered at Macy's, in case you're curious. As far as I can tell, nobody bought them the Cuisinart food processor, so feel free to whip out that credit card.)

Whatever doubts you might have about the rapid progress of the science of resurrection, just try sharing those doubts with Michelle

Funk, now Michelle York. Try telling her that the past two decades of her life shouldn't have happened. Try telling her that she shouldn't need that Cuisinart because she shouldn't be alive. Good luck with that.

But if we want to celebrate Michelle Funk's wedding, we also need to be ready to face the results of stories that don't end well. We need to be ready to care for patients like Joe, the first patient I ever brought back to life. He spent eighteen days in the ICU because I was so quick to start CPR. And we need to be able to support his family who endured eighteen days of watching, waiting, and making increasingly difficult choices about whether and when to stop treatment.

Confronting these results will be challenging because we're still not very good at the "soft" side of medicine that helps people make choices about which treatments they want. We're bad at supporting families who get stuck making difficult decisions as Joe's family had to. And we're really awful at saying no to treatment that won't help us. So although I'm hopeful about the future and about what's going to be possible someday, I won't be truly optimistic about our ability to take care of hearts and brains until we get much better at taking care of people.

ACKNOWLEDGMENTS

This hasn't been the sort of book that one writes in solitude, comfortably parked on the front porch of a cabin in the mountains. Writing—and researching—has required considerable travel, correspondence, telephone conversations, and the help of dozens of people. For instance, there were many researchers who took time out of their hectic schedules to talk with me. Among those, I owe special thanks to David Gaieski, Joshua Lampe, Lance Becker, Cheng Chi Lee, Hannah Carey, and Peter Klopfer. I'm also grateful to others who made connections and introductions, like Andy Kofke, Ed Dickenson, Sam Tisherman, John Nilsson, Suzannah Hughes, and Greg Marok. Along the way, Lauren Mancuso was extremely helpful in digging up obscure facts about resuscitation, many of which I deeply regret were just too bizarre to print. I'm also grateful to the animals that helped move this research along, including Penny the horse, Petunia the pig, Chucky the squirrel, a mouse known to his close friends as #0011, numerous squirrels, groundhogs, pigs, monkeys, "mongrel dogs of medium size," and of course all of the sleeping members of Team Lemur.

The folks at Alcor were particularly helpful (and trusting) in letting a skeptic into their midst. Despite my challenging questions about the science of cryonics, those cryonauts were uniformly open and welcoming. They probably won't make that mistake again.

This book was also possible because I'm fortunate to work with wonderful people at the University of Pennsylvania who are supportive, encouraging, and tolerant of the demands of my writing. They are also generous in working around my schedule when I venture out into the wide world to meet hibernating ground squirrels, cryonauts, and other anomalies of nature. So I've been able to write this book in large part thanks to the flexibility of people like Joan Doyle, PJ Brennan, and Zeke Emanuel, who don't seem to mind when I skip a meeting to chase lemurs. I've also been able to embark on these expeditions because I'm fortunate to work with people who keep things running smoothly in my absence, like Laura Bender, Meredith Dougherty, Sue Foster, Sue Kristiniak, Amy Corcoran, Barbara Reville, Nina O'Connor, and Beth Reimet.

On the writing side, my agent, Chris Bucci, at Ann McDermid & Associates, provided much-needed encouragement and guidance as this book was taking shape. He handed that role seamlessly to my editor, Niki Papadopoulos, who has been a welcome source of boundless enthusiasm and constructive editorial advice.

Finally, I'm also grateful to the patients and others who were willing to share their stories. To protect their confidentiality, I've described them using pseudonyms only. But they know who they are.

NOTES

1: The Big Mac Rule of Resuscitation and the Search for the Limits of Life

3 The full story: The Michelle Funk case report comes from RG Bolte et al. (1988) "The use of extracorporeal rewarming in a child submerged for 66 minutes," *Journal of the American Medical Association* 260(3): 377–79.

4 In fact, the conventional wisdom: JP Orlowski (1987) "Drowning, near-drowning, and ice-water submersions," *Pediatric Clinics of North America* 34(1): 75–92.

5 an editorial accompanying that article: JP Orlowski (1988) "Drowning, near-drowning, and ice-water drowning," *Journal of the American Medical Association* 260(3): 390–91.

9 "When hateful old age": HG Evelyn-White, trans., "Homeric Hymn to Aphrodite," theoi.com/Text/HomericHymns3.html [accessed January 10, 2014].

2: Why Amsterdam Used to Be a Good Place to Commit Suicide

13 Anne Wortman is lying facedown: The Wortman case report comes from the translated notes of the Amsterdam Society. Dr. Thomas Cogan, trans., *Memoirs of the Society Instituted at Amsterdam in Favour of Drowned Persons:*

For the Years 1767, 1768, 1769, 1770, and 1771 (London: Printed for G Robinson, Pater-Noster Row, 1773) vol. 27; chapter XV.

18 **"hospitals and public charities":** Ibid.

18 **"To blow into the intestines" . . . "the warm irritating fumes":** These suggestions regarding the inflation of the intestines come from Cogan, *Memoirs.*

19 **"Yet in vain is it condemned" . . . "The populace will not easily renounce":** Ibid., chapter VII.

19–20 **"Let one of the assistants, applying his mouth" . . . "Since no body can affirm with certainty":** Ibid.

21 **"a neat structure":** *Illustrated London News,* August 19, 1844, p. 144.

23 **"within a reasonable time of immersion":** Accounts differ about just how long these early pioneers thought someone could be dead before being brought back to life. See, for instance, en.wikipedia.org/wiki/Royal_Humane_Society. However, in the first annual report, Hawes specifies two hours as the upper limit. W. Hawes, *Transactions of the Royal Humane Society; Dedicated by Permission to His Majesty by W. Hawes. Volume 1* (Eighteenth Century Editions Online; Print Editions. 2013; originally published 1794), p. 11.

23 **two each brought fifteen citizens:** *The Forty-Eighth Annual Report of the Royal Humane Society for the Recovery of Persons Apparently Drowned or Dead* (London: Printed for the Society and to be had at the Society's House, 29 Bridge-Street, Blackfriars, 1882).

24 **"a father to the fatherless":** Royal Humane Society, "History," royalhumanesociety.org.uk/html/history.html [accessed January 10, 2014].

25 **Of these, some of the strongest recommendations:** *Annual Report of the Royal Humane Society for the Recovery of Persons Apparently Dead.* Seventieth Annual Report. London: Compton and Ritchie, 1844.

25 **its seventieth annual report:** Ibid.

25 **"For some time" . . . "the metallic trachea tube":** Ibid.

26 **However, the Society produced:** This helpful advice is provided by Charles Dickens in his *Dictionary of the Thames, from its Source to the Nore,* and attributed to the Royal Humane Society (London: MacMillan & Co.), p. 57.

28 **Actually, I learned later from an authoritative source:** Mickey Eisenberg, *Life in the Balance: Emergency Medicine and the Quest to Reverse Sudden Death* (Oxford: Oxford University Press, 1997), 63. However, the Society Web site indicates that it was demolished in 1954: Royal Humane Society, "History," royalhumanesociety.org.uk/html/history.html [accessed January 10, 2014]. Personally, I prefer to believe that the Germans viewed the house's stash of feathers as a weapons stockpile that needed to be destroyed.

40 William Tebb, an English businessman: William Tebb and Edward Perry Vollum, *Premature Burial and How It May Be Prevented, with Special Reference to Trance, Catalepsy, and Other Forms of Suspended Animation* (London: Swan, 1905).

41 "Charles Walker was supposed to have died": Tebb, *Premature Burial,* 101.

41 "ALMOST every intelligent and observant person": Ibid., 98.

42 the case of Anne Green: Mickey Eisenberg's sleuthing unearthed the following example and he highlights it as the first clear case of a near miss: JT Hughes (1982) "Miraculous deliverance of Anne Green: An Oxford case of resuscitation in the 17th century," *British Medical Journal* 285: 1792–793.

3: The Ice Woman Meets the Strange New Science of Resuscitation

45 When she was twenty-nine years old: "Frozen Woman: A 'Walking Miracle,'" February 11, 2009, cbsnews.com/2100-18564_162-156476.html [accessed January 10, 2014].

47 "I think it's amazing that I'm alive": Ibid.

57 "However . . . after the experiment": Peter Christian Abildgaard, *Tentamina electrica in animalibus, Inst Soc Med Havn*. 1775; 2:157-61. This source and its translation come from two physicians, Dean Jenkins and Stephen Gerred, whose tireless work to teach the art of reading EKGs is exceeded only by their passion for the topic's history. And by their inexhaustible sense of humor regarding experiments like this one. See: "A (not so) brief history of electrocardiography," ecglibrary.com/ecghist.html [accessed January 10, 2014].

57 Birds seem to have borne: Charles Schechter, *Exploring the Origins of Electrical Cardiac Stimulation* (Minneapolis: Medtronic, 1883), 361. I found this titillating reference to Humboldt in Mickey Eisenberg, *Life in the Balance* (Oxford: Oxford University Press, 1997).

58 But the physician and historian Mickey Eisenberg: Eisenberg's history is, for my money, about the best one there is. It helps enormously that Eisenberg himself is a highly respected researcher in this field, and that his account is eminently readable. See Eisenberg, *Life in the Balance*.

58 The next decade saw: Charles Kite, "An Essay on the Recovery of the Apparently Dead," *Annual Report 1788: Humane Society* (London: C. Dilly, 1788), 225–44. A more accessible and detailed description of Kite's work can be found in AG Alzaga, J Varon, and P Baskett (2005) "The resuscitation greats. Charles Kite: The clinical epidemiology of sudden cardiac death and the origin of the early defibrillator," *Resuscitation* 64(1): 7–12.

59 The next big breakthrough came in 1947: Eisenberg, *Life in the Balance*, 188.

60 As Beck was ending the operation: CS Beck, WH Pritchard, and HS Feil (1947) "Ventricular fibrillation of long duration abolished by electric shock," *Journal of the American Medical Association* vol. 135: 1230–233.

74 "It is well known that human beings": The experiments on groundhogs, monkeys, and "mongrel dogs of medium size" described here are from: WG Bigelow and JE McBirnie (1953) "Further experiences with hypothermia for intracardiac surgery in monkeys and groundhogs" *Annals of Surgery* 137(3): 361–65.

74 "I was interested in cooling": D Bigelow (1997) "Dr. Wilfred Gordon Bigelow named to Canadian Medical Hall of Fame," *The Forge: The Bigelow Society Quarterly*, 26(4): 68.

74 dogs could survive fifteen minutes: WG Bigelow, JC Callaghan, and JA Hopps (1950) "General hypothermia for experimental intracardiac surgery; the use of electrophrenic respirations, an artificial pacemaker for cardiac standstill and radio-frequency rewarming in general hypothermia," *Annals of Surgery* 132: 531–39.

74 This is a success rate: Bigelow and McBirnie, "Further experiences," p. 365.

75 "more akin to man" . . . "He appears normal": Ibid., p. 361.

76 "We feel, however, that cooling dogs": WG Bigelow, WK Lindsay, and WF Greenwood (1950) "Hypothermia; its possible role in cardiac surgery: an investigation of factors governing survival in dogs at low body temperatures," *Annals of Surgery* 132: 849–66.

78 The Safar Center did several experiments: There are many published articles resulting from these experiments. A few are listed here: A Nozari et al. (2006) "Critical time window for intra-arrest cooling with cold saline flush in a dog model of cardiopulmonary resuscitation," *Circulation* 113: 2690–696; X Wu et al. (2008) "Emergency preservation and resuscitation with profound hypothermia, oxygen, and glucose allows reliable neurological recovery after 3 h of cardiac arrest from rapid exsanguination in dogs," *Journal of Cerebral Blood Flow & Metabolism* 28: 302–11; C Kovner et al. (2006) "Mild hypothermia during prolonged cardiopulmonary cerebral resuscitation increases conscious survival in dogs," *Critical Care Medicine* 32: 2110–116; A Nozari et al. (2004) "Suspended animation can allow survival without brain damage after traumatic exsanguination cardiac arrest of 60 minutes in dogs," *Journal of Trauma-Injury Infection & Critical Care* 57: 1266–275; W Behringer et al. (2003) "Survival without brain damage after clinical death of 60–120 mins in dogs using suspended animation by profound hypothermia," *Critical Care Medicine* 31: 1523–531.

78 "custom-bred, male hunting dogs": A Nozari et al., "Critical time window," p. 267.

79 News reports, for instance: The quotes that follow are from "Blood Swapping Reanimates Dead Dogs," June 25, 2005, foxnews.com/story/0,2933,160903,00.html [accessed January 10, 2014].

83 They found a 16 to 25 percent absolute increase: The Hypothermia After Cardiac Arrest Study Group (2002) "Mild therapeutic hypothermia to improve the neurologic outcome after cardiac arrest," *New England Journal of Medicine* 346: 549–56; SA Bernard et al (2002) "Treatment of comatose survivors of out-of-hospital cardiac arrest with induced hypothermia," *New England Journal of Medicine* 346: 557–63.

83 In one large study: J Nielsen et al. (2009) "Outcome, timing, and adverse events in therapeutic hypothermia after out-of-hospital cardiac arrest," *Acta Anaesthesiologica Scandinavica* 53: 926–34.

4: Science Fiction, Space Travel, and Suspended Animation

87 On October 7, 2006: BBC News, "Japanese Man in Mystery Survival," December 21, 2006, news.bbc.co.uk/2/hi/asia-pacific/6197339.stm [accessed January 10, 2014].

89 "First known human case": Justin McCurry, "Injured Hiker Survived 24 Days on Mountain by 'Hibernating,'" *Guardian,* December 20, 2006, guardian.co.uk/world/2006/dec/21/japan.topstories3 [accessed January 10, 2014].

92 "decline the trouble of migration": Aristotle (350 BCE), *Historiae Animalium* [The history of animals] D'Arcy Wentworth Thompson, trans., book VIII, part 16, classics.mit.edu/Aristotle/history_anim.8.viii.html [accessed January 10, 2014].

92 "If you prick out the eyes of swallow chicks": Ibid., book VI, part 5, classics.mit.edu/Aristotle/history_anim.6.vi.html.

93 "That the golden hamster": CP Lyman and PO Chatfield (1955) "Physiology of hibernation in mammals," *Physiological Reviews* 35(2): 403–25.

93 "Concentration of urine": ML Zatzman and FE South (1975) "Concentration of urine by the hibernating marmot," *American Journal of Physiology* 228(5): 1336–340.

93 "The unique maturation response": WA Wimsatt and FC Kallen (1957) "The unique maturation response of the graafian follicles of hibernating vespertilionid bats and the question of its significance," *Anatomical Record* 129(1): 115–31.

93 This area of research got a boost: F Smith and MM Grenan (1951) "Effect of hibernation upon survival time following whole-body irradiation in the marmot (*Marmota monax*)," *Science* 113(2946): 686–88.

97 When scientists first began to examine: BR Landau and AR Dawe (1958) "Respiration in the hibernation of the 13-lined ground squirrel," *American Journal of Physiology* 194(1): 75–82.

98 Specifically, researchers focused on an area of the brain: E Satinoff (1965) "Impaired recovery from hypothermia after anterior hypothalamic lesions in hibernators," *Science* 148(3668): 399–400.

99 During the winter of 1967–68: AR Dawe and WA Spurrier (1969) "Hibernation induced in ground squirrels by blood transfusion," *Science* 163: 298–99.

101 Alas, subsequent attempts to replicate: B Abbotts, LC Wang, and JD Glass (1979) "Absence of evidence for a hibernation 'trigger' in blood dialyzate of Richardson's ground squirrel," *Cryobiology* 16: 179–83.

101 They act as a class of opioid receptors: CV Borlongan et al. (2009) "Hibernation-like state induced by an opioid peptide protects against experimental stroke," *BMC Biology* 7: 31.

102 To find out, Willis gathered: JS Willis (1964) "Potassium and sodium content of tissues of hamsters and ground squirrels during hibernation," *Science* 146(3643): 546–47.

103 Anti-apoptotic enzymes, which go by awkward names: C Fleck and HV Carey (2005) "Modulation of apoptotic pathways in intestinal mucosa during hibernation," *American Journal of Physiology: Regulatory Integrative and Comparative Physiol*ogy 289: R586–95.

104 And they produced more bile: SL Lindell, SL Klahn, TM Piazza, MJ Mangino, JR Torrealba, JH Southard, et al. (2005) "Natural resistance to liver cold ischemia-reperfusion injury associated with the hibernation phenotype." *American Journal of Physiology Gastrointestinal Liver Physiology* 288: G473–80.

105 researchers removed the hearts of a bunch of rabbits: SF Bolling et al. (1997) "Use of 'natural' hibernation induction triggers for myocardial protection," *Annals of Thoracic Surgery* 64(3): 623–27.

107 The term "artificial hibernation": S Simpson and PT Herring (1905) *Journal of Physiology* 32: 305.

107 The challenges of maintaining body temperature: H Laborit (1954) "General technic of artificial hibernation," *International Record of Medicine & General Practice Clinics* 167(6): 324–27.

108–10 "So obvious are the advantages" . . . "One formed the impression": JW Dundee et al. (1953) "Hypothermia with autonomic block in man," *British Medical Journal* 2: 1237–43.

115 Like Thomas, they didn't: A Percy et al. (2009) "Deep hypothermic circulatory arrest in patients with high cognitive needs: full preservation of cognitive abilities," *Annals of Thoracic Surgery* 87: 117–23.

119 **In 2005, though:** KH Dausmann, J Glos, JU Ganzhorn, G Heldmaier (2005) "Hibernation in the tropics: lessons from a primate," *Journal of Comparative Physiology B* 175: 147–55.

123 **Finally, and perhaps most interestingly:** E Jerlhag et al. (2009) "Requirement of central ghrelin signaling for alcohol reward," *Proceedings of the National Academy of Sciences of the United States of America* 106(27): 11318–1323.

123 **There's even very limited evidence that:** M Hotta et al. (2009) "Ghrelin increases hunger and food intake in patients with restricting-type anorexia nervosa: a pilot study," *Endocrine Journal* 56(9): 1119–128.

125 **To understand what was in that injection:** IS Daniels et al. (2010) "A role of erythrocytes in adenosine monophosphate initiation of hypometabolism in mammals," *Journal of Biological Chemistry* 285: 20716–0723.

126 **So constant darkness seems to create:** J Zhang et al. (2006) "Constant darkness is a circadian metabolic signal in mammals," *Nature* 439(7074): 340–43.

128 **He pointed out that when hibernating bears exhale:** H Clapp (1868) "Notes of a fur hunter" *American Naturalist* 1: 653.

128 **In 2010, Oivind Toien and his colleagues:** O Toien et al. (2011) "Black bears: independence of metabolic suppression from temperature," *Science* 331: 906–09.

5: The Deep-Freeze Future: Cryonauts Venture to the Frontiers of Immortality

141 **Instead, it allows itself to freeze:** Much of the early research to understand how frogs manage to freeze was done by the husband-and-wife team of Ken and Janet Storey. See: KB Storey and JM Storey (1984) "Biochemical adaption for freezing tolerance of the wood frog, *Rana sylvatica*," *Journal of Comparative Physiology B* 155: 29–36.

142 **One of the first reports of successful cryopreservation:** C Polge, AU Smith, and AS Parkes (1949) "Revival of spermatozoa after vitrification and dehydration at low temperatures," *Nature* 164: 666.

144 **I've come to the fortieth anniversary conference:** Alcor Life Extension Foundation, "About Alcor: Our History," alcor.org [accessed January 10, 2014].

146 **The slim, petite woman:** "Suspended Animation," suspendedinc .com [accessed January 10, 2014].

148 **Consider the sad case of Alcor patient 113:** "Alcor's 113th Patient," alcor. org/blog/alcors-113th-patient/ [accessed January 23, 2014].

149 Inspired by those rumors: Robert Ettinger, "The Penultimate Trump," *Startling Stories* 17, March 1948.

154 As the weather gets colder: Storey and Storey, "Biochemical adaption."

154 What is even more interesting: JM Storey and KB Storey (1985) "Triggering of cryoprotectant synthesis by the initiation of ice nucleation in the freeze tolerant frog, *Rana sylvatica*," *Journal of Comparative Physiology B* 156: 191–95.

156 These so-called antifreeze proteins: G Amir et al. (2004) "Prolonged 24-hour subzero preservation of heterotopically transplanted rat hearts using antifreeze proteins derived from arctic fish," *Annals of Thoracic Surgery* 77: 1648–655.

156 The same team reported later: A Elami et al. (2008) "Successful restoration of function of frozen and thawed isolated rat hearts," *Journal of Thoracic Cardiovascular Surgery* 135: 666–72.

156 Fahy mentions a study of his own: GM Fahy et al. (2009) "Physical and biological aspects of renal vitrification," *Organogenesis* 5(3): 167–75.

160 A case in point, which is either a sign of progress or a cautionary tale: Mike Darwin, "Dear Dr. Bedford," alcor.org/Library/html/BedfordLetter.htm [accessed January 10, 2014].

160 Bedford was moved within days: Mike Darwin, "Evaluation of the Condition of Dr. James H. Bedford After 24 Years of Cryonic Suspension" alcor.org/Library/html/BedfordCondition.html [accessed January 10, 2014].

161 So, how well did he do? Ibid. [accessed February 5, 2014]. All of the gruesome descriptions of poor James Bedford that follow are taken from this report.

6: Crowdsourcing Survival

176 We don't know everything that happened: feed://radio.foxnews.com/tag/tracey-halvorson/feed/ [accessed February 4, 2014].

177 "Staff at Senior Living Home Refuses": CBS News, "Staff at Senior Living Home Refuses to Perform CPR on Dying Woman," March 1, 2013, losangeles.cbslocal.com/2013/03/01/staff-at-senior-living-home-refuse-to-perform-cpr-on-dying-woman [accessed February 5, 2014].

178 "I think anyone with any clinical training": Dana Edelson, quoted in Judith Graham, "Amid CPR Controversy, Many Unanswered Questions," The New Old Age Blog, *New York Times,* March 6, 2013, newoldage.blogs.nytimes.com/2013/03/06/amid-cpr-controversy-many-unanswered-questions [accessed February 5, 2014].

178 "All of us . . . have a duty to respond": Dale Jamieson, quoted in Ibid.

178 So she truly was a bystander: KGET, "Glenwood Gardens in the National Spotlight," March 4, 2013, kget.com/news/local/story/Glenwood-Gardens-in-the-national-spotlight/uxvuCpj170y6xQDz0bsNQQ.cspx [accessed February 5, 2014].

179 Although it's difficult to get accurate estimates: AL Valderrama et al. (2011) "Cardiac arrest patients in the emergency department—National Hospital Ambulatory Medical Care Survey, 2001–2007," *Resuscitation* 82(10): 1298–301.

179 Add to that the cardiac arrests that happen in hospitals: RM Merchant et al. (2011) "Incidence of treated cardiac arrest in hospitalized patients in the United States," *Critical Care Medicine* 39(11): 2401–406.

180 "He sparked me into a lifelong pursuit": P Safar, "On the future of reanimatology" (2000) *Academic Emergency Medicine* 7(1): 75–89.

180 "In four breaths": Mickey Eisenberg, *Life in the Balance*, 89–90.

181 The boy survived: JO Elam, "Rediscovery of expired air methods for emergency ventilation," in *Advances in Cardiopulmonary Resuscitation*, Peter Safar, ed. (New York: Springer Verlag, 1977), 263–65.

181 Inspired by that success: JO Elam, ES Brown, and JD Elder (1954) "Artificial respiration by mouth-to-mask method; a study of the respiratory gas exchange of paralyzed patients ventilated by operator's expired air," *New England Journal of Medicine* 250(18): 749–54.

181 Finally, when the laypeople: P Safar (1958) "Ventilatory efficacy of mouth-to-mouth artificial respiration: Airway obstruction during manual and mouth-to-mouth artificial respiration," *Journal of the American Medical Association* 167: 335–41.

182 Somewhere in the middle of an experiment: WB Kouwenhoven, JR Jude, and GG Knickerbocker (1960) "Closed-chest cardiac massage," *Journal of the American Medical Association* 173: 94–7.

182 She lived: Eisenberg, *Life in the Balance*, 124–25.

183 After contemplating the mask he'd discovered: Ibid., 102.

186 In case you, too, want some fun bedtime reading: RA Berg et al. (2010) "Part 5: Adult Basic Life Support, 2010 American Heart Association Guidelines for Cardiopulmonary Resuscitation and Emergency Cardiovascular Care," *Circulation* 122: S685–705.

187 The more compressions you do: B Abella et al. (2005) "Chest compression rates during cardiopulmonary resuscitation are suboptimal: a prospective study during in-hospital cardiac arrest," *Circulation* 111: 428–34.

188 In one meta-analysis of adults: M Hupfl, HF Selig, and P Nagele (2010) "Chest-compression-only versus standard cardiopulmonary resuscitation: a meta-analysis," *Lancet* 376(9752): 1552–557.

188 That's important because: I Stiell et al. (2012) "What is the role of chest compression depth during out-of-hospital cardiac arrest resuscitation?" *Critical Care Medicine* 40: 1192–198.

189 fatigue results in a measureable decline: S Manders and F Geijsel (2009) "Alternating providers during continuous chest compressions for cardiac arrest: every minute or every two minutes?" *Resuscitation* 80: 1015–018.

194 Jose Antonio Adams and Paul Kurlansky: miamiheartresearch .org/pgzmotion/references.html [accessed February 4, 2014].

195 And in one scene, as Tristin recalls it: Ben Forer, "Arizona 9-Year-Old Boy, Tristin Saghin, Saved Sister with CPR, Congratulated by Movie Producer Jerry Bruckheimer," April 22, 2011, abcnews.go.com/Health/ arizona-year-boy-tristin-saghin -saved-sister-cpr/story?id=13428007 [accessed February 5, 2014].

195 one study found that bystander-initiated CPR: C Sasson et al. (2012) "Association of neighborhood characteristics with bystander-initiated CPR," *New England Journal of Medicine* 367(17): 1607–15.

196 There are many reasons why bystanders: C Sasson et al. (2013) "Increasing cardiopulmonary resuscitation provision in communities with low bystander cardiopulmonary resuscitation rates: a science advisory from the American Heart Association for healthcare providers, policymakers, public health departments, and community leaders," *Circulation* 127: 342–50; R Swor et al. (2006) "CPR training and CPR performance: do CPR-trained bystanders perform CPR?" *Academy of Emergency Medicine* 13(6): 596–601; C Vaillancourt, IG Stiell, and GA Wells (2009) "Understanding and improving low bystander CPR rates: a systematic review of the literature," *Canadian Journal of Emergency Medicine* 10(1): 51–65.

196 What they found was that 75 percent of patients who underwent CPR: SJ Diem, JD Lantos, and JA Tulsky (1996) "Cardiopulmonary resuscitation on television: miracles and misinformation," *New England Journal of Medicine* 334(24): 1578–582.

200 Indeed they do: ML Weisfeldt et al. (2011) "Survival after application of automatic external defibrillators before arrival of the emergency medical system: evaluation in the resuscitation outcomes consortium population of 21 million," *Journal of the American College of Cardiology* 55: 1713–720; VL Roger et al. (2011) "Heart disease and stroke statistics—2011 update: a report from the American Heart Association," *Circulation* 123: e18–e209.

200 And it turns out that, for some reason: ML Weisfeldt et al. (2011) "Ventricular tachyarrhythmias after cardiac arrest in public versus at home," *New England Journal of Medicine* 364(4): 313–21.

202 Dr. Raina Merchant, a colleague of mine at Penn: R Merchant et al. (2012) "Locating AEDs in an urban city: A geospatial view (Abstract)," *Circulation* 126: A58.

203 more likely to have someone start CPR if you die in public: SL Caffrey et al. (2002) "Public use of automated external defibrillators," *New England Journal of Medicine* 347(16): 1242–247; TD Valenzuela et al. (2000) "Outcomes of rapid defibrillation by security officers after cardiac arrest in casinos," *New England Journal of Medicine* 343(17): 1206–209; RA Swor et al. "Cardiac arrest in private locations: different strategies are needed to improve outcome," *Resuscitation* 58(2): 171–76.

204–5 In fact, it may be that a majority of patients: AF Hernandez et al. (2007) "Sex and racial differences in the use of implantable cardioverter-defibrillators among patients hospitalized with heart failure," *Journal of the American Medical Association* 298(13): 1525–532.

206 Implantation rates rose in almost all: HG Mond and A Proclemer (2009) "The 11th world survey of cardiac pacing and implantable cardioverter-defibrillators: calendar year 2009—a World Society of Arrhythmia's project," *Pacing and Clinical Electrophysiology* 34(8): 1013–27.

206 In one study of patients with heart failure: GH Bardy et al. (2005) "Amiodarone or an implantable cardioverter-defibrillator for congestive heart failure," *New England Journal of Medicine* 352(3): 225–37.

206 Still, it's possible that ICDs: AJ Moss et al. (2001) "Survival benefit with an implanted defibrillator in relation to mortality risk in chronic coronary heart disease," *American Journal of Cardiology* 88(5): 516–20.

206 Estimates are hard to come by: PW Groeneveld et al. (2006) "Costs and quality-of-life effects of implantable cardioverter-defibrillators," *American Journal of Cardiology* 98(10): 1409–415.

207 Although, interestingly: Ibid.

208 patients who receive a shock within the first year: DL Carroll and GA Hamilton (2005) "Quality of life in implanted cardioverter-defibrillator recipients: the impact of a device shock," *Heart Lung* 34(3): 169–78.

208 symptoms of post-traumatic stress disorder: KH Ladwig et al. (2008) "Posttraumatic stress symptoms and predicted mortality in patients with implantable cardioverter-defibrillators: results from the prospective living with an implanted cardioverter-defibrillator study," *Archives of General Psychiatry* 65(11): 1324–330.

209–10 Despite the availability of trained medical staff: MN Shah, RJ Fairbanks, and EB Lerner (2007) "Cardiac arrests in skilled nursing facilities: continuing room for improvement?" *Journal of the American Medical Directors' Association* 8: e27–31.

221 Conservatively speaking, the price tag for the ambulance ride: G Nichol et al. (2009) "Cost-effectiveness of lay responder defibrillation for out-of-hospital cardiac arrest," *Annals of Emergency Medicine* 54: 226–35 e1-2.

7: When Is "Dead" Really Dead? Listen for the Violins.

225 "As Coroner, I must aver": *The Wizard of Oz* (1939); Victor Fleming, director.

227 His team observed an average drop in temperature: E Blackstone, M Morrison, and M Roth (2005) "H_2S induces a suspended animation-like state in mice," *Science* 308: 518.

227 Pigs, apparently, don't respond: J Li et al. (2008) "Effect of inhaled hydrogen sulfide on metabolic responses in anesthetized, paralyzed, and mechanically ventilated piglets," *Pediatric Critical Care Medicine* 9: 110–12.

227 Nor do sheep: P Haouzi et al. (2008) "H_2S induced hypometabolism in mice is missing in sedated sheep," *Respiratory Physiology and Neurobiology* 160: 109–15.

228 That trial began in May 2009: M Roth (2010) "Reduction in ischemia-reperfusion mediated cardiac injury in subjects undergoing coronary artery bypass graft surgery." http://clinicaltrials.gov/ct2/show/NCT00858936 [accessed February 5, 2014].

228 A query to Ikaria: Samina Bari, e-mail response to author from Ikaria, January 16, 2013.

INDEX

Read on for the first chapter from *Stoned*,
available now in hardcover.

1

The Blacksmith and the Boxer

The mobile home has traveled many hard miles. Like old luggage, its tan exterior is scuffed and faded and marred by countless dents. In the gusty wind blowing across the surrounding field on the outskirts of Denver, its frame rocks restlessly on bald tires, making it seem alive. The windows are obscured by curtains. A poster on the door shows the outline of a human-shaped shooting target, pocked with bullet holes. The caption below it reads "There's nothing in here worth dying for."

I turn to the man standing next to me. Nathan Pollack is in his seventies, with snow-white hair and a whorled white beard that ends in a sharp point at his chin. His black beret and rumpled tweed sport coat make him look like he could be a professor at a small liberal arts college. He is in fact a hospice doctor, like me, and he's about to introduce me to a patient whose story I very much want to hear.

I knock, and the door opens a few inches, revealing the head of an enormous dog with bulbous eyes and small, clipped ears. His intimidation potential is elevated a bit by the fact that his teeth are at the level of my throat.

It turns out that the dog—all eighty pounds of him—just wants someone to scratch behind his ears. Still, I let Pollack go in first.

Inside, the mobile home smells of stale marijuana smoke, dog, old socks, and compost. The interior is dark despite the bright winter day outside. Several heavy quilts cover the windows as insulation against the biting wind, and another covers the doorway that leads to the cab in front. The cozy womblike space is crammed with a bed, two small chairs, and a tiny kitchen area. There's faded wallpaper and a mock-Tudor ceiling constructed of stucco interspersed with faux wood beams. It's oddly homey.

As my eyes adjust to the dark, I get a good look at the man Pollack has brought me to meet. Caleb* is in his forties, lean and wiry. His neatly trimmed mustache is at odds with brown hair that's long and boyishly unkempt. He's sitting cross-legged on his bed, warmly dressed in a thermal long-sleeved shirt, a flannel shirt, jeans, and well-worn athletic socks.

Caleb's hands are shaking, and I recognize the resting tremor of Parkinson's disease. He tells me that he got that diagnosis several years ago and thinks the unusually early onset is probably a result of his work as a blacksmith and a welder. "You inhale every heavy metal known to man," he says. "No surprise that they do all kinds of damage to your brain."

I ask him his dog's name, and Caleb's face contorts into what might pass for a smile.

"Rocky."

Of course. What else would you call a boxer? I laugh, and Caleb does, too. He clutches his left side in obvious pain. Then that passes, and he starts to tell me the story I came to hear.

Two years ago, he was diagnosed with rectal cancer. He received what he described as "aggressive" treatment in Wisconsin. "Some days I didn't think I could take any more," he says. "But then I'd go by the children's hospital and I'd watch those kids playing outside and I'd think, 'If they can handle this, then I can, too.' "

* "Caleb" isn't his real name. Here and throughout, I've used first-name pseudonyms for all of the patients I've met, as well as for anyone else who might get into legal trouble for their marijuana-related activities. The people for whom I've included full (real) names gave me their permission to do so and spoke on the record.

His doctors told him that treatment could prolong his life, but wouldn't cure him. So he suffered through as much treatment as he could stand, and then he drove west to Colorado where he knew he'd be able to get marijuana legally to treat his pain and control his nausea.

Along the way, he also discovered that marijuana didn't only relieve those symptoms.

"It also keeps me from being an asshole." He pauses. "Well, too much of an asshole, anyway. It sort of blunts the sharp edges. I'm not so pissed off all the time about everything—at everyone."

"Having cancer sucks," he admits. "But when you've got your red card"—that's Colorado's medical marijuana card—"dying sucks a little less."

It turned out, however, that getting marijuana wasn't so easy.

As soon as he parked his mobile home in this suburb outside of Denver, Caleb found Pollack's hospice, which promised free marijuana to its patients. That's essential, because Caleb lives on about $600 a month. But Pollack's hospice had stopped providing marijuana because the management didn't want to run afoul of federal laws that still classify marijuana as an illegal substance. So once he was in Denver, Caleb wasn't able to get marijuana from Pollack's hospice, and he couldn't afford to buy his own.

As Caleb tells his story, he drops the courtly politeness with which he'd greeted me a few minutes ago, and his language becomes increasingly laced with profanity. He goes on to say how much he depends on the marijuana he smokes. But he can't afford it.

"Oh, there's plenty of places to buy. That's the hell of it. Sure it's legal, but . . ." He rubs his thumb and forefinger together in the universal sign for money. "You have to be able to pay."

I ask him if he'd found another hospice that could give him marijuana, and he grins.

"No, I didn't really look." He winks at Pollack. "I'll complain, sure. I'm not shy. Especially when I'm pissed off. But these are good folks. They take good care of me."

He's stayed with Pollack's hospice, and he's gotten marijuana from so-called angels in the community who make donations directly to him. That keeps the hospice out of trouble, which is fine with Caleb.

"I wouldn't want anybody going to jail for me."

Besides, he has a plan.

"I'm growing my own. See?" He points.

To my left, at the rear of the mobile home, there's a grow light suspended over a plastic tub. The light sways gently in the wind, casting undulating shadows of what looks like about thirty immature marijuana plants. A few are about eighteen inches tall and look like they might start budding in a couple of weeks. But most are no more than a couple of inches high.

Caleb admits that those smaller plants were the result of an accident. Literally. He went to pick up plants from someone who was moving and wanted to give them a good home. But problems ensued.

"I'm not the best driver, you know?"

I look back at the psychoactive garden. I nod.

Caleb is a little vague about details, but apparently while he was driving, the microwave fell off its shelf and several plants lost their lives. He salvaged what he could by taking cuttings and creating clones, which are the Lilliputian plants that take up most of his little greenhouse.

I notice a box on a shelf behind him. It's labeled COMFORT KIT.

This is a set of medications that many hospices provide to their patients in case of emergencies. It generally includes morphine for pain, and benzodiazepines like Valium for anxiety and seizures.

Has he used them?

"Hell no. They don't work. Why would I waste my time with them?" He gestures at a dime bag of marijuana on the kitchen counter. "That's all I need, right there."

Caleb recognizes the irony of this arrangement. His hospice can provide a variety of drugs such as morphine free of charge. These are drugs that have the potential to cause a fatal overdose, and which can be addictive. But the drug that really helps him—marijuana—is out of reach.

"Those are your tax dollars at work, man. You're paying for the government to spend money on that box of shitty drugs that I'm not going to touch. That's a waste. But what I really need, they can't give me. Does that make sense?"

I admit that it doesn't. As if in agreement, Rocky the boxer hops off the bed where he'd been sitting with Caleb and rests his chin on my leg.

Does Medical Marijuana "Work"?

In Pollack's Toyota, heading back to Denver, I think about Caleb's predicament. I'm a little shell-shocked by what I've heard. I'd always thought of medical marijuana as a joke. Or a "treatment" that would always be described in just that way, hemmed in by ironic quotes.

Yet Caleb is dying and in pain. And he wants it—needs it—for relief. This isn't a guy who wants to get high for fun. This is a man who has led a hard life and who doesn't want to suffer more than he has to.

Some patients I've taken care of in my work as a hospice and palliative care doctor have admitted to me that they use marijuana for symptoms like pain or nausea. (And I'm pretty sure that for every patient who has been honest with me about using marijuana, there are many more who haven't.) They've made it into a joke, just as I have.

Laughing uncomfortably, one Vietnam veteran told me he started smoking half a joint per day more than twenty years ago to control his PTSD symptoms. Better to be high than out of his head, right? I laughed, too, and we moved on to talk about his cancer pain that I was treating.

I'm pretty sure I shouldn't have laughed. Like Caleb, that veteran needed help. He'd found something that relieved his symptoms, and he needed my support.

I don't remember what I said. Probably some version of what all doctors say when confronted with alternative medicine we don't understand (or, in the case of marijuana, common but usually illegal). "Well, I guess it can't hurt." Or, "If it makes you feel better . . ." But I didn't ask him whether it helped him, or how.

Marijuana is the *only* thing that's helped Caleb. This is a guy who is letting $100 worth of morphine and other interesting drugs sit in his closet. He'd rather have a joint.

What's more, Caleb turned to marijuana to avoid drugs like morphine. Not only is he convinced that marijuana is helpful, he's convinced that it's better than the legal drugs he can get for free. It might be safer, too.

Does marijuana "work"?

When I met Caleb in early 2014, the debate about legalizing medical marijuana was just beginning to get national attention. But that debate was about ethics and morality. Do people have a right to use it? And how should laws be crafted?

No one was really talking seriously about whether marijuana has any medical value. Or whether it's safe. Or, setting the bar a little lower, whether it's at least as safe as other drugs that doctors prescribe.

My goal with this book is to shift the national discussion, and to bring these questions about effectiveness and safety to the forefront. These are the questions we really need to be asking. And these are the questions that I wanted to be able to answer in order to decide, as a physician, what advice I should give my patients when they ask me whether they should use medical marijuana.

In this book I've attempted to look carefully—and critically—at the evidence. I searched for high-quality published studies, and I interviewed researchers who are doing those studies. In short, I subjected medical marijuana to the same scrutiny that I'd give to the drugs that the pharmaceutical industry tries to sell to physicians like me.

First, I investigated whether marijuana works as medicine. I scoured medical journals, interviewed patients, and visited physicians and researchers. I wanted to find out what we know about marijuana's effect on conditions related to the brain, such as insomnia and PTSD and seizures. I wanted to learn whether it could be useful in treating physical symptoms such as nausea and weight loss, and whether the ingredients in marijuana might actually treat diseases like dementia, multiple sclerosis, or cancer.

Then I explored how people get marijuana into their systems. That might seem like a silly question, but it's not an easy one to answer. I learned how to infuse the ingredients of marijuana into beer and wine and brownies and ointment. I discovered how effective joints and bongs are at delivering marijuana's ingredients to our brains compared to newer technologies such as vaporizers. Along the way, I learned how to make hash. I even discovered what I've been told is the world's best pot brownie recipe. (It's on page 261, in case you want to skip ahead.)

All that is great, but is marijuana *safe*? For most readers of this book, this might be the most important question. So I sifted through the evidence about whether long-term use causes brain damage, or mental illness, or addiction. I discovered there are other risks of marijuana use beyond its effects on the brain—from infertility to shrinking penises and everything in between. I'll also share my own misguided experiences with medical marijuana, and I'll take you for a drive with one of my patients shortly after he smoked a joint.

Next, how are patients like Caleb figuring out whether marijuana could help them? How do they sort through the risks and potential benefits of this stuff? Where do they turn for advice? And how reliable is that advice?

Finally, what does the future hold for medical marijuana? Is it on its way to being the next superdrug? Should it be legalized and made more available? Or is this all merely hype to justify recreational drug use?

These are high-stakes questions. As of this writing, twenty-three states and the District of Columbia have legalized marijuana for medical purposes, and a few have allowed it for recreational purposes, too. It seems very likely that medical marijuana will be legal in most states in the not-too-distant future. And in states where it's legal, lots of people are using it. According to one study in California, 5 percent of people surveyed had used medical marijuana.[1] That means that my questions about marijuana's effectiveness, and especially its safety, have big implications for hundreds of thousands of people.

These numbers are likely to increase further as more dispensaries open to sell marijuana. That was the conclusion of another study in

California that surveyed more than eight thousand people in fifty cities and mapped their proximity to marijuana dispensaries. The results: people who were close to a dispensary used more marijuana.[2]

The states that have legalized it have done so with very broad criteria. You can use it for chronic pain, anxiety, insomnia, arthritis, and a wide variety of other conditions. Most people who are embracing it are not near the end of life. They are students, construction workers, police officers, teachers, and—I can attest—at least one doctor.

If medical marijuana works, then its growing popularity is good news. But if it doesn't work, then it's an enormous waste of time and money. Even worse, if marijuana isn't safe, we'll have an enormous public health crisis on our hands. As you'll see later in the book, if lots of people are using medical marijuana, even a very small risk could result in lots of people being harmed or even killed.

As I started writing this book, the public was just beginning to recognize those risks. For instance, one study found that long-term marijuana use was associated with brain atrophy—in other words, brain shrinkage.[3] But shortly after that rather alarming study made headlines, *The New York Times* published an editorial in favor of legalizing medical marijuana, and discounting its risks. So is it safe or not?[4]

Whether marijuana is safe and effective is particularly important because its safety will determine how it's regulated. For much of its recent history, marijuana has been classified in the United States as a Schedule I substance. This category is reserved for drugs like heroin that are believed to have significant risks but no known medical benefits. As long as marijuana continues to live in that category, it's going to be impossible to create national rules about how it can be used, prescribed, and distributed. It's also very difficult for researchers to do studies that can help us figure out whether and how it works. And patients like Caleb won't be able to get it from a hospice.

I wasn't sure where I'd come out on the other end of this journey. Pro or con? I'm also not sure which side you'll be on when you reach the last chapter. But no matter what your verdict, I think I can promise you an interesting journey that will give you enough information to make up your own mind about whether marijuana could help you.

Finally, a book about pot wouldn't be complete without some caveats.

First, a word about names: Throughout this book, I'll be referring to "marijuana," over the objections of some purists who prefer the term *cannabis*, as in *Cannabis sativa*. They correctly note that the name *marijuana* (actually *marihuana*, imported from Mexico) was bestowed by narcotics enforcement officials back in the 1940s as a way to scare off potential users. Back then, the connotations of foreignness and crime (think *Reefer Madness*) were enough to deter a few people. (Of course, those connotations also probably made the stuff irresistible to others.) Although I admit that cannabis is the proper term, marijuana is so widely recognized that I've chosen to use this name instead, with all due apologies to terminological purists.

Second, you shouldn't rely on this book as your sole source of medical advice. If you want to learn how to treat symptoms like pain, you should also see a physician.

However, this book will help you decide whether medical marijuana might help you. I've written it for people like my own patients, and like Caleb, who are struggling with distressing symptoms and looking for help. It's intended to be a summary of the sort of advice I would want them to have when they're deciding whether medical marijuana might help them.

For instance, I'll tell you how medical marijuana can be helpful, and what symptoms it's most effective in treating. I'll also warn you about marijuana's risks, and how you can avoid them. Along the way, I'll tell you enough about the science of medical marijuana to understand how it works, and what it might be able to do for us in the future.